Complex Analysis for Practical Engineering

Kozo Sato

Complex Analysis for Practical Engineering

 Springer

Kozo Sato
University of Tokyo
Tokyo
Japan

ISBN 978-3-319-35180-3   ISBN 978-3-319-13063-7   (eBook)
DOI 10.1007/978-3-319-13063-7

Springer Cham Heidelberg New York Dordrecht London

Printed on acid-free paper

Springer International Publishing AG Switzerland is part of Springer Science+Business Media
(www.springer.com)

Dedicated to M.S. and S.S.

Preface

"A lot of subjects to study; no spare time for unnecessary themes." For busy learners, necessity is the mother of motivation. The subject symbolized by

$$i^2 = -1$$

may appear to them as a leisured manipulation in an imaginary world, which is a wasteful misconception. Complex analysis is of practical use in a real world.

This book is devoted to demonstrating the practical utility of complex analysis. Unlike standard mathematical textbooks, theoretical explanations do not come first. Instead, the individual chapters start with motivating problems which help the readers understand the need to learn the relevant topics and the application procedures of such topics to practical engineering problems.

Although the book covers complex analysis, mainly in the context of potential flow problems, the basic concept and application methodologies are general and can be easily extended to other engineering problems, including diffusion, heat conduction, and gravitational and electrostatic fields. Rather than a reference book, this is a book for engineering practitioners.

Looking out over the forest of maples in the sunset, the ninth century poet Tu Mu composed "View from the Cliffs," which concludes with the following phrase:

The frosted leaves are more brilliant than any flowers of spring

The seemingly impractical mathematics, complex analysis, exhibits its utility as brilliantly as other seemingly practical mathematics.

Tokyo, October 2014   Kozo Sato

Contents

Chapter 1
Potential Theory in Practical Engineering

Motivating Problem 1: Groundwater Contamination

Groundwater flows in the x direction (eastward direction) with a uniform flow velocity $q_u/A = 1$, where A is the cross-sectional area of the flow medium. It turns out that contaminants flow into the porous medium at the location $(-1, 0)$ with a flow rate per unit thickness $q_s/h = 6$, where h is the thickness of the medium. There is a residential area in the northeast, and to examine the contamination, a monitoring well is deployed at the location $(1.5, 1.5)$. Figure 1.1 illustrates the situation.

Task 1-1 Draw the pressure profile in the flow domain and predict whether the contaminants are detected at the monitoring well.

Task 1-2 Draw the velocity profile in the flow domain and predict whether the contaminants are detected at the monitoring well.

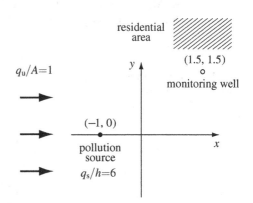

Fig. 1.1 Uniform flow in the x direction with a pollution source at $(-1, 0)$ and a monitoring well at $(1.5, 1.5)$

residential area

$q_u/A=1$

y

$(1.5, 1.5)$
o
monitoring well

$(-1, 0)$

pollution source

$q_s/h=6$

x

© Springer International Publishing Switzerland 2015
K. Sato, *Complex Analysis for Practical Engineering*,
DOI 10.1007/978-3-319-13063-7_1

• Solution Strategy to Motivating Problem 1
The first step in solving engineering problems is to derive the governing equation that describes the physical process of interest. For Motivating Problem 1, mathematical modeling of groundwater flow in porous media needs to be established.

Two types of flow take place simultaneously: uniform flow and flow induced by a pollution source, to which mathematical solutions need to be individually obtained. Then the final step is to find a way to combine individual solutions into a complete solution, with which the corresponding pressure and velocity profiles are obtained.

1.1 Potential Flow

With the assumptions appropriate to the physical process of interest, potential flow of a fluid is introduced. Laplace's equation, a well-known partial differential equation, is derived in the context of flow problems.

1.1.1 Continuity Equation

Let us consider a control volume of a rectangular parallelepiped box of dimensions Δx, Δy, and Δz, as shown in Fig. 1.2, inside a flow field. A fluid enters and leaves the box through its surfaces. The law of mass conservation states that there is no net change in the mass of a fluid in the representative control volume.

The mass flow rate into the element in the x direction is the product of the fluid density ρ, the fluid velocity V_x, and the surface area $\Delta y \Delta z$, that is, $\rho V_x \Delta y \Delta z$. The mass flow rate out of the element may change as the fluid passes through the box. Let this change be $\Delta(\rho V_x)\Delta y \Delta z$; then the mass flow rate leaving the opposite surface of the box in the x direction is given by $(\rho V_x + \Delta(\rho V_x))\Delta y \Delta z$. The excess of inflow

Fig. 1.2 Mass fluxes across surfaces of a control volume in a three-dimensional flow field

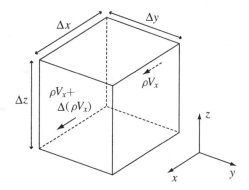

over outflow of mass in the x direction during a short time interval Δt (between t and $t + \Delta t$) is given by $-\Delta(\rho V_x)\Delta y \Delta z \Delta t$.

In a similar way, the mass flow rates into and out of the element in the y and z directions are obtained. For the control volume, the excess of inflow over outflow of mass during a short time interval Δt is obtained as

[Mass inflow] $-$ [Mass outflow]

$$= -\left[\Delta(\rho V_x)\Delta y \Delta z + \Delta(\rho V_y)\Delta z \Delta x + \Delta(\rho V_z)\Delta x \Delta y\right]\Delta t \qquad (1.1)$$

For the same control volume, let the mass in the box at time t be $\rho\Delta x\Delta y\Delta z$ and at time $t + \Delta t$ be $(\rho + \Delta\rho)\Delta x\Delta y\Delta z$. Then, the change of mass during Δt in the box is given by

$$[\text{Mass accumulation}] = \Delta\rho\Delta x\Delta y\Delta z \qquad (1.2)$$

By the law of mass conservation, the excess of mass given by Eq. 1.1 must be equal to the mass accumulation given by Eq. 1.2. Dividing the resultant equation by $\Delta x\Delta y\Delta z\Delta t$ yields

$$\frac{\Delta\rho}{\Delta t} + \frac{\Delta(\rho V_x)}{\Delta x} + \frac{\Delta(\rho V_y)}{\Delta y} + \frac{\Delta(\rho V_z)}{\Delta z} = 0 \qquad (1.3)$$

and letting Δx, Δy, Δz, and Δt approach zero gives

$$\frac{\partial\rho}{\partial t} + \frac{\partial(\rho V_x)}{\partial x} + \frac{\partial(\rho V_y)}{\partial y} + \frac{\partial(\rho V_z)}{\partial z} = 0 \qquad (1.4)$$

From Eq. A.3, Eq. 1.4 is expressed with $\mathbf{V} = (V_x, V_y, V_z)$ as

$$\frac{\partial\rho}{\partial t} + \nabla \cdot \rho\mathbf{V} = 0 \qquad (1.5)$$

which is the continuity equation for flow of a compressible fluid.

If a fluid is incompressible, that is, $\rho = $ constant, Eq. 1.5 reduces to

$$\nabla \cdot \mathbf{V} = 0 \qquad (1.6)$$

In addition, when the flow behavior in the xy plane is the same as in any plane parallel to the xy plane, it is sufficient to consider the flow of a sheet of fluid in two-dimensional coordinates. Then Eq. 1.6 is written as

$$\frac{\partial V_x}{\partial x} + \frac{\partial V_y}{\partial y} = 0 \qquad (1.7)$$

which is the continuity equation for flow of an incompressible fluid in two-dimensional Cartesian coordinates.

The assumption of two-dimensional flow is adopted and the mathematical development and applications in subsequent parts of this book are restricted to two dimensions.

1.1.2 Irrotationality

Let us consider a control element of a rectangle of dimensions Δx and Δy, as shown in Fig. 1.3, inside a two-dimensional flow field. For the element under rotation during a short time interval Δt, the points B and C move perpendicular to the flow in the x and y directions, respectively.

Let V_y be the velocity in the y direction at A and $V_y + \Delta V_y$ be the velocity at B. The excess of the velocity at B over that at A is ΔV_y. During Δt, the point B moves to B' (relative to A) and the length BB' is given by $\Delta V_y \Delta t$. Consequently, the line segment AB rotates by an angle $\Delta \delta$, and for a small value of $\Delta \delta$, it follows that

$$\Delta \delta = \frac{BB'}{AB} = \frac{\Delta V_y}{\Delta x} \Delta t \tag{1.8}$$

The corresponding angular velocity ω_{AB} is given by $\Delta \delta / \Delta t$ and letting Δt and Δx approach zero yields

$$\omega_{AB} = \lim_{\substack{\Delta t \to 0 \\ \Delta x \to 0}} \frac{\Delta \delta}{\Delta t} = \frac{\partial V_y}{\partial x} \tag{1.9}$$

In a similar way, the angular velocity of the line segment AC is given by

$$\omega_{AC} = -\frac{\partial V_x}{\partial y} \tag{1.10}$$

where the negative sign is introduced so that the counterclockwise direction is taken to be positive. The total angular velocity ω is defined as the average of the angular

Fig. 1.3 Rotational motion of a control element in a two-dimensional flow field

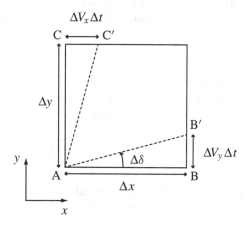

velocities ω_{AB} and ω_{AC}:

$$\omega = \frac{\omega_{AB} + \omega_{AC}}{2} = \frac{1}{2}\left(\frac{\partial V_y}{\partial x} - \frac{\partial V_x}{\partial y}\right) \tag{1.11}$$

which is called the rotation.

If viscous stresses are negligible, there is no mechanism to induce rotation. When there is no rotation at each point in a flow domain, the flow is said to be irrotational. Hence, the condition of irrotationality is given by

$$\frac{\partial V_y}{\partial x} - \frac{\partial V_x}{\partial y} = 0 \tag{1.12}$$

the left-hand side of which is called the vorticity. From Eq. A.5, Eq. 1.12 is equivalent to

$$\nabla \times \mathbf{V} = \mathbf{0} \tag{1.13}$$

which is also the condition of irrationality.

1.1.3 Discharge and Circulation

Let C be a smooth curve given by the vector $\mathbf{r}(s) = x(s)\mathbf{i_x} + y(s)\mathbf{i_y}$, where s is the arc length of C and $\mathbf{i_x}$ and $\mathbf{i_y}$ are the unit vectors in the x and y directions, respectively. Then the unit tangent and normal vectors, $\mathbf{u}$ and $\mathbf{n}$, are respectively given by

$$\begin{cases} \mathbf{u} = \dfrac{d\mathbf{r}}{ds} = \dfrac{dx}{ds}\mathbf{i_x} + \dfrac{dy}{ds}\mathbf{i_y} \\[2mm] \mathbf{n} = \dfrac{dy}{ds}\mathbf{i_x} - \dfrac{dx}{ds}\mathbf{i_y} \end{cases} \tag{1.14}$$

which satisfy the orthogonality, $\mathbf{u} \cdot \mathbf{n} = 0$.

For flow with the velocity vector $\mathbf{V} = (V_x, V_y)$, the normal component of velocity on C is given by $\mathbf{V} \cdot \mathbf{n}$. The discharge across C is defined by

$$Q = \int_C \mathbf{V} \cdot \mathbf{n} ds = \int_C \left(V_x dy - V_y dx\right) \tag{1.15}$$

where Q is the discharge in terms of volume per unit thickness per unit time. When C is a closed curve, from Green's theorem (Appendix B.1), the flux is given by a double integral as

$$Q = \oint_C \left(V_x dy - V_y dx\right) = \iint_R \left(\frac{\partial V_x}{\partial x} + \frac{\partial V_y}{\partial y}\right) dx dy \tag{1.16}$$

where R is the area bounded by C. If the fluid is incompressible, the discharge across a closed curve C is zero:

$$Q = 0 \tag{1.17}$$

which is the condition of incompressibility and implies that the integrand in the double integral in Eq. 1.16 vanishes at every point in the flow domain, that is,

$$\frac{\partial V_x}{\partial x} + \frac{\partial V_y}{\partial y} = 0 \tag{1.18}$$

which is equivalent to the continuity equation, Eq. 1.7.

In a similar way, the tangential component of velocity on C is given by $\mathbf{V} \cdot \mathbf{u}$, and the circulation Γ along C is defined by

$$\Gamma = \int_C \mathbf{V} \cdot \mathbf{u} ds = \int_C \left(V_x dx + V_y dy \right) \tag{1.19}$$

When C is a closed curve, from Green's theorem, the circulation is given by a double integral as

$$\Gamma = \oint_C \left(V_x dx + V_y dy \right) = \iint_R \left(\frac{\partial V_y}{\partial x} - \frac{\partial V_x}{\partial y} \right) dx dy \tag{1.20}$$

If the circulation along a closed curve C is zero

$$\Gamma = 0 \tag{1.21}$$

the integrand in the double integral in Eq. 1.20 vanishes at every point in the flow domain, that is,

$$\frac{\partial V_y}{\partial x} - \frac{\partial V_x}{\partial y} = 0 \tag{1.22}$$

which is equivalent to the condition of irrotationality, Eq. 1.12.

1.1.4 Laplace's Equation

When the flow is irrotational, the circulation Γ for any closed curve in the flow domain is zero. From Eq. 1.19 with $\mathbf{u} ds = \mathbf{dr}$, it follows that

$$\Gamma = \oint_C \mathbf{V} \cdot \mathbf{dr} = 0 \tag{1.23}$$

A vector function $\mathbf{V}$ satisfying Eq. 1.23 is said to be conservative. From vector calculus (Appendix C), Eq. 1.23 is equivalent to the existence of a scalar function Φ such that[1]

$$\mathbf{V} = -\nabla\Phi \qquad (1.24)$$

where the scalar function Φ is called the velocity potential. Since the velocity vector $\mathbf{V}$ is the gradient of velocity potential, irrotational flow is also referred to as potential flow.

In the case of an incompressible fluid, the continuity equation at Eq. 1.6 must be satisfied, and it follows that

$$\nabla \cdot \mathbf{V} = \nabla \cdot (-\nabla\Phi) = 0 \qquad (1.25)$$

which reduces to Laplace's equation:

$$\nabla^2\Phi = 0 \qquad (1.26)$$

For potential flow of an incompressible fluid, the velocity potential Φ satisfies Laplace's equation.

In Cartesian coordinates, Eq. 1.24 gives the components of the velocity vector as

$$\begin{cases} V_x = -\dfrac{\partial\Phi}{\partial x} \\[2mm] V_y = -\dfrac{\partial\Phi}{\partial y} \end{cases} \qquad (1.27)$$

and Eq. 1.26 is written as

$$\frac{\partial^2\Phi}{\partial x^2} + \frac{\partial^2\Phi}{\partial y^2} = 0 \qquad (1.28)$$

for flow in two dimensions.

1.1.5 Flow in Porous Media

The mass balance equation for saturated flow of a compressible fluid in a porous medium follows from a similar development as described in Sect. 1.1.1. It can be shown that the volumetric porosity ϕ, a fraction of the volume of voids over the

---

[1] The negative sign in Eq. 1.24 is in accordance with the sign convention adopted in this book; that is, a fluid flows from a point with a higher potential to another point with a lower potential.

total volume, is equal to the average areal porosity, and thus, the continuity equation
Eq. 1.5 is modified for flow in porous media as

$$\frac{\partial(\rho\phi)}{\partial t} + \nabla \cdot \rho\phi\mathbf{V_s} = 0 \tag{1.29}$$

where $\mathbf{V_s}$ is the seepage velocity vector representing the rate of flow volume per unit
pore area normal to the flow (also known as average linear velocity, linear velocity,
pore velocity, and interstitial velocity).

Assuming that a fluid is incompressible and denoting $\mathbf{V} = \phi\mathbf{V_s}$ reduces Eq. 1.29 to

$$\nabla \cdot \mathbf{V} = 0 \tag{1.30}$$

where $\mathbf{V}$ is the Darcy velocity vector representing the rate of flow volume per unit
total area normal to the flow ($\mathbf{q}/A$).

According to Darcy's law, the Darcy velocity $\mathbf{V}$ is proportional to the pressure
gradient ∇p in the direction of flow, proportional to the permeability k of the medium
and inversely proportional to the viscosity μ of the fluid, that is,

$$\mathbf{V} = \frac{\mathbf{q}}{A} = -\frac{k}{\mu}\nabla p \tag{1.31}$$

The negative sign is added because ∇p is negative in the direction of flow.[2]

When the flow takes place through a homogeneous isotropic medium and the fluid
viscosity is constant, k/μ is a constant scalar. Let us define the velocity potential Φ
for flow in porous media as

$$\Phi = \frac{k}{\mu}p \tag{1.32}$$

Then Darcy's law can be rewritten as

$$\mathbf{V} = -\nabla\Phi \tag{1.33}$$

Substituting Eq. 1.33 into the continuity equation (Eq. 1.30) yields Laplace's
equation:

$$\nabla^2\Phi = 0 \tag{1.34}$$

For flow of an incompressible fluid in porous media, the velocity potential Φ satisfies
Laplace's equation.

---

[2] A fluid flows from a point with a higher pressure to another point with a lower pressure, which is
consistent with the sign convention used in Eq. 1.24.

1.2 Physical Processes in Potential Theory

Laplace's equation is the most representative of the class of elliptic partial differential equations. The theory of Laplace's equation is called potential theory, which is applied not only to potential flow but also to many engineering problems, including but not limited to Fickian diffusion, heat conduction, gravitational fields, and electrostatic fields. Although potential flow is mainly considered in this book, the mathematical techniques to be developed in the following chapters are also applicable to these problems.

1.2.1 Fickian Diffusion

As the difference in velocity potential induces fluid flow, the difference in solute concentration induces solute mass flow. This phenomenon is expressed by Fick's law of diffusion as

$$\mathbf{U_m} = -D\nabla C \tag{1.35}$$

where $\mathbf{U_m}$ is the mass flux, C the mass concentration, and D the diffusion coefficient.

From the law of mass conservation, the following continuity equation in terms of C holds:

$$\frac{\partial C}{\partial t} + \nabla \cdot \mathbf{U_m} = 0 \tag{1.36}$$

Substituting Eq. 1.35 into Eq. 1.36 and assuming that D is constant yields

$$\frac{\partial C}{\partial t} - D\nabla^2 C = 0 \tag{1.37}$$

which, under steady-state conditions, reduces to

$$\nabla^2 C = 0 \tag{1.38}$$

For Fickian diffusion under steady-state conditions, the mass concentration C satisfies Laplace's equation.

1.2.2 Heat Conduction

Heat conduction is the transfer of heat energy induced by the difference in temperature. This phenomenon is expressed by Fourier's law as

$$\mathbf{U_h} = -\kappa \nabla T \tag{1.39}$$

where $\mathbf{U_h}$ is the heat flux, T the temperature, and κ the thermal conductivity.

From the law of conservation of energy, it follows that

$$\rho c \frac{\partial T}{\partial t} + \nabla \cdot \mathbf{U_h} = 0 \tag{1.40}$$

where ρ is the density and c the specific heat. Substituting Eq. 1.39 into Eq. 1.40 and assuming that κ is constant yields

$$\rho c \frac{\partial T}{\partial t} - \kappa \nabla^2 T = 0 \tag{1.41}$$

which, under steady-state conditions, reduces to

$$\nabla^2 T = 0 \tag{1.42}$$

For heat conduction under steady-state conditions, the temperature T satisfies Laplace's equation.

1.2.3 Gravitational Fields

A gravitational field is a force vector field caused by the attraction between massive objects. Since the gravitational field $\mathbf{F}$ is known to be conservative (Appendix C), for any closed curve C, it follows that

$$\oint_C \mathbf{F} \cdot \mathbf{dr} = 0 \tag{1.43}$$

and there exists a scalar function Φ_g such that

$$\mathbf{F} = -\nabla \Phi_g \tag{1.44}$$

where Φ_g is the gravitational potential.

Gauss's law for gravitational fields relates the distribution of mass to the resulting gravitational field as

$$\nabla \cdot \mathbf{F} = -4\pi G \rho \tag{1.45}$$

where ρ is the density and G the gravitational constant. Substituting Eq. 1.44 into Eq. 1.45 yields

$$\nabla^2 \Phi_g = 4\pi G \rho \tag{1.46}$$

In empty space $\rho = 0$, and it follows that

$$\nabla^2 \Phi_g = 0 \tag{1.47}$$

In empty gravitational fields, the gravitational potential Φ_g satisfies Laplace's equation.

1.2.4 Electrostatic Fields

An electrostatic field is a force vector field caused by the attraction or repulsion between charged bodies. Since the electrostatic field $\mathbf{E}$ is known to be conservative (Appendix C), for any closed curve C, it follows that

$$\oint_C \mathbf{E} \cdot \mathbf{dr} = 0 \tag{1.48}$$

and there exists a scalar function Φ_e such that

$$\mathbf{E} = -\nabla \Phi_e \tag{1.49}$$

where Φ_e is the electrostatic potential.

Gauss's law for electrostatic fields relates the distribution of electric charge to the resulting electric field as

$$\nabla \cdot \varepsilon_0 \mathbf{E} = \rho_e \tag{1.50}$$

where ρ_e is the electric charge density and ε_0 the permittivity. Substituting Eq. 1.49 into Eq. 1.50 and assuming that ε_0 is constant yields

$$\nabla^2 \Phi_e = -\frac{\rho_e}{\varepsilon_0} \tag{1.51}$$

If the charge distribution is located outside the domain of interest $\rho_e = 0$, it follows that

$$\nabla^2 \Phi_e = 0 \tag{1.52}$$

In electrostatic fields without ρ_e, the electrostatic potential Φ_e satisfies Laplace's equation.

1.3 Velocity Potential

In this section, the fundamental equations describing the physical processes appearing in Motivating Problem 1 are developed. A powerful tool, the principle of superposition, is also introduced to combine individual processes into one.

1.3.1 Uniform Flow

Uniform flow occurs when a fluid flows in a single direction (in the x direction, for instance) and the cross-sectional area A to flow is constant. For incompressible fluids, the flow rate q_u and thus the fluid velocity $V_x = q_u/A$ are constant at all points along the flow path.

The behavior of uniform flow in the x direction can be expressed by the following differential form of Darcy's law (Eq. 1.33):

$$V_x = \frac{q_u}{A} = -\frac{d\Phi}{dx} \tag{1.53}$$

Separating the variables and integrating between two lengths x_0 and x gives

$$\int_{x_0}^{x} V_x dx = \frac{q_u}{A} \int_{x_0}^{x} dx = -\int_{\Phi_0}^{\Phi(x)} d\Phi \tag{1.54}$$

which results in

$$\Phi(x) = -\frac{q_u}{A}x + \Phi_0 \tag{1.55}$$

where Φ_0 is the value at x_0 and for simplicity x_0 is set to 0 without essential loss of generality. The negative sign in Eq. 1.55 indicates that the velocity potential decreases in the x direction (in the flow direction).

1.3.2 Sources and Sinks

Sources and sinks are the points in the two-dimensional flow plane at which fluids emerge and disappear, respectively. Flow paths are radial lines emanating from a source and converging toward a sink. Unlike uniform flow, the cross-sectional area A across which flow occurs is not constant. At radius r in the flow plane of thickness h, the cross-sectional area is the surface area of a cylinder, that is, $A = 2\pi r h$. For flow of an incompressible fluid with a constant flow rate of q_w, the fluid velocity in the radial direction $V_r = q_w/A$ is a function of r.

The behavior of radial flow in the r direction can be expressed by the following differential form of Darcy's law (Eq. 1.33):

$$V_r = \frac{q_w}{2\pi r h} = -\frac{d\Phi}{dr} \tag{1.56}$$

Separating the variables and integrating between two radii r_0 and r gives

$$\int_{r_0}^{r} V_r dr = \frac{q_w}{2\pi h} \int_{r_0}^{r} \frac{dr}{r} = -\int_{\Phi_0}^{\Phi(r)} d\Phi \tag{1.57}$$

which results in

$$\Phi(r) = -\frac{q_w}{2\pi h} \ln r + \Phi_0 \tag{1.58}$$

where Φ_0 is the value at r_0 and for simplicity r_0 is set to 1 without essential loss of generality. When q_w is positive, Eq. 1.58 indicates that the velocity potential decreases in the r direction, which corresponds to emanating flow from a source. With a negative value of q_w, Eq. 1.58 gives converging flow toward a sink.

The location of a source or sink is arbitrary. In general, the radius r in Cartesian coordinates is given by

$$r = \sqrt{(x - x_w)^2 + (y - y_w)^2} \tag{1.59}$$

where (x_w, y_w) is the location of a source or sink.

Example 1.1 Let us confirm that the velocity potentials satisfy Laplace's equation. For uniform flow, from Eq. 1.55, it follows that

$$\nabla^2 \Phi = \frac{\partial^2 \Phi}{\partial x^2} = -\frac{\partial}{\partial x} \frac{q_u}{A} = 0$$

For a source or sink, from Eq. 1.58 and using Eq. A.12, it follows that

$$\nabla^2 \Phi = \frac{1}{r} \frac{\partial}{\partial r} \left(r \frac{\partial \Phi}{\partial r} \right) = -\frac{1}{r} \frac{\partial}{\partial r} \frac{q_w}{2\pi h} = 0$$

The velocity potential for uniform flow and that for a source or sink indeed satisfy Laplace's equation.

1.3.3 Equipotential Lines

If $\Phi(x, y)$ is the velocity potential at a point (x, y) in the xy plane, then the level curves of Φ are called the equipotential lines. Every point in an equipotential line is at the same velocity potential. As the contour lines on a topographical map show the ground elevations, equipotential lines show the variation of velocity potential throughout a flow domain.

A plot of equipotential lines is obtained by setting Φ equal to a constant in the equation

$$\Phi = \Phi(x, y) = \text{constant} \tag{1.60}$$

which describes a family of curves, for various values of the constant.

Example 1.2 Figure 1.4 shows the equipotential lines of uniform flow given by Eq. 1.55 and a source or sink given by Eq. 1.58. For uniform flow, the equipotential lines are straight lines parallel to the y axis and they are all the same distance apart. For a source or sink, the equipotential lines are concentric circles. These circles become further apart with increasing radius r, and, conversely, they get crowded in the vicinity of the source or sink point, which implies a rapid change in velocity potential as the source or sink is approached.

Since a fluid flows from a point with a higher velocity potential to another point with a lower potential, a flow path crosses the equipotential lines and its direction can be derived. For uniform flow, the flow paths are straight lines in the x direction and for a source or sink, the paths are emanating from or converging toward the source or sink point.

(a)   **(b)**

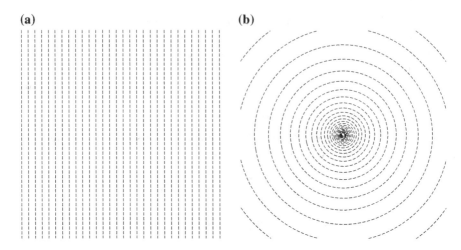

Fig. 1.4 Equipotential lines. **a** Uniform flow. **b** Source or sink

1.3.4 Principle of Superposition

Complicated physical processes may be modeled with simpler processes using the principle of superposition. This principle states that if multiple different functions Φ_j are solutions of linear differential equation $L(\Phi_j) = 0$, where L represents a linear operator, then any linear combination

$$\Phi = \sum_{j=1}^{n} c_j \Phi_j \tag{1.61}$$

where c_j are constants, is also a solution of $L(\Phi) = 0$. The constants c_j are adjusted so that the boundary conditions are also satisfied.

In the case of Laplace's equation $L = \nabla^2$, it follows that

$$\sum_{j=1}^{n} c_j \nabla^2 \Phi_j = \sum_{j=1}^{n} c_j \left[\frac{\partial^2 \Phi_j}{\partial x^2} + \frac{\partial^2 \Phi_j}{\partial y^2} \right] = 0 \tag{1.62}$$

Since c_j are constants, Eq. 1.62 can be rewritten as

$$\sum_{j=1}^{n} c_j \left[\frac{\partial^2 \Phi_j}{\partial x^2} + \frac{\partial^2 \Phi_j}{\partial y^2} \right] = \frac{\partial^2 \sum_{j=1}^{n} c_j \Phi_j}{\partial x^2} + \frac{\partial^2 \sum_{j=1}^{n} c_j \Phi_j}{\partial y^2} = 0 \tag{1.63}$$

Substituting Eq. 1.61 into Eq. 1.63 yields Laplace's equation in terms of Φ, as stated in the principle of superposition:

$$\frac{\partial^2 \Phi}{\partial x^2} + \frac{\partial^2 \Phi}{\partial y^2} = 0 \tag{1.64}$$

By applying the principle of superposition, the net response caused by multiple flow processes can be obtained as the sum of responses that would have been caused by each flow individually. For instance, uniform flow with a source, as considered in Motivating Problem 1, can be modeled by addition of the solution for uniform flow alone and that for a source alone.

• Solution to Task 1-1
By use of superposition, the velocity potential for uniform flow with a source located at (x_s, y_s) can be obtained by addition of Eq. 1.55 and Eq. 1.58 as

$$\Phi(x, y) = -\frac{q_u}{A}x - \frac{q_s}{4\pi h}\ln\left[(x - x_s)^2 + (y - y_s)^2\right] \tag{1.65}$$

where Φ_0 in Eq. 1.55 and Eq. 1.58 are set to zero.
Substituting $q_u/A = 1$, $q_s/h = 6$, and $(x_s, y_s) = (-1, 0)$ yields

$$\Phi(x, y) = -x - \frac{3}{2\pi}\ln\left[(x + 1)^2 + y^2\right] \tag{1.66}$$

with which the corresponding velocity potentials are computed and the equipotential lines are obtained as shown in Fig. 1.5.

In the vicinity of the source at $(-1, 0)$, the equipotential lines are approximately concentric circles and getting crowded, which indicates a dominant influence of the source on the flow profile and, consequently, a rapid increase in pressure as the

Fig. 1.5 Equipotential lines for uniform flow in the x direction with a source at $(-1, 0)$

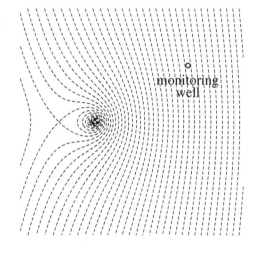

monitoring well

source point is approached. In contrast, away from the source, the equipotential lines are approximately parallel straight lines which are orthogonal to the flow direction, implying that an influence of the source is attenuated.

At the monitoring well, a moderate influence of the source and a dominant influence of uniform flow are observed. However, the pressure profile does not provide sufficient information to predict whether the contaminants are detected at the monitoring well.

● **Solution to Task 1-2**

The Darcy velocity vector is given by Eq. 1.33. For uniform flow with a source, from Eq. 1.66, it follows that

$$\begin{cases} V_x = -\dfrac{\partial \Phi}{\partial x} = 1 + \dfrac{3}{\pi}\dfrac{x+1}{(x+1)^2 + y^2} \\ V_y = -\dfrac{\partial \Phi}{\partial y} = \dfrac{3}{\pi}\dfrac{y}{(x+1)^2 + y^2} \end{cases}$$

with which the corresponding velocity vectors are computed and the velocity profile is obtained as shown in Fig. 1.6.

The velocity vector field describes the local direction and magnitude of the Darcy velocity. The velocity vectors in the vicinity of the source at $(-1, 0)$ are in the radially outward direction and exhibit larger values in magnitude. Away from the source, the velocity vectors are approximately in the x direction, which reflects uniform flow.

Fig. 1.6 Velocity vectors for uniform flow in the x direction with a source at $(-1, 0)$

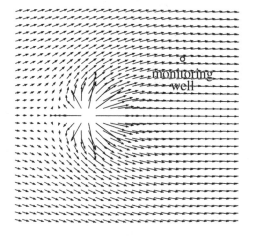

As is the case with the pressure profile in Task 1-1, a moderate influence of the source and a dominant influence of uniform flow are observed at the monitoring well. Although the velocity profile provides a rough image of flow paths, the information is still not enough to predict whether the contaminants are detected at the monitoring well.

Chapter 2
Complex Potential and Differentiation

Motivating Problem 2: Extraction of Contaminated Groundwater

Let us revisit Motivating Problem 1. If the contamination is observed at the monitoring well, an extraction well is drilled at the location $(1, 0)$, as shown in Fig. 2.1, to withdraw contaminated groundwater. For the extraction system to be optimized, a precise evaluation of flow behavior of contaminants is essential.

Task 2-1   Draw the flow paths in the flow domain and predict whether the contaminants are detected at the monitoring well.

Task 2-2   Draw the flow paths in the flow domain and evaluate the minimum pumping rate q_e/h at the extraction well to avoid the contamination at the monitoring well.

Task 2-3   Redo Task 2-2 in an analytical manner.

Fig. 2.1 Uniform flow in the x direction with a pollution source at $(-1, 0)$, a monitoring well at $(1.5, 1.5)$, and an extraction well at $(1, 0)$

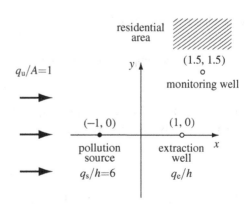

© Springer International Publishing Switzerland 2015
K. Sato, *Complex Analysis for Practical Engineering*,
DOI 10.1007/978-3-319-13063-7_2

• **Solution Strategy to Motivating Problem 2**

Although the velocity potential (or its derivative) provides a rough image of flow profiles, as seen in Motivating Problem 1, it does not convey sufficient information to draw exact flow paths. To evaluate the direction of flow, another mathematical function must be acquired.

The strategy, therefore, is to derive a new function that conveys such information. Of course, the new function must be consistent with the physical nature of the velocity potential. This derivation process reveals the reason why complex variables need to be introduced and the usefulness of complex analysis in practical engineering.

2.1 Complex Numbers

To perform complex analysis, the real number system needs to be extended to the complex number system. The algebraic properties and geometric representation of complex numbers are discussed.

2.1.1 Definition

There is no real number x that satisfies the equation $x^2 = -1$. To manipulate these types of equations, the set of complex numbers is introduced. By definition, a complex number z is an ordered pair (x, y) of real numbers x and y, written as

$$z = (x, y) \tag{2.1}$$

The ordered pair $(0, 1)$ is called the imaginary unit and is denoted by i as

$$i = (0, 1) \tag{2.2}$$

which has the property of

$$i^2 = -1 \tag{2.3}$$

In practice, a complex number z is expressed in the form

$$z = x + iy \tag{2.4}$$

The real numbers x and y are referred to as the real part of z and the imaginary part of z, respectively, and are written as

$$\begin{cases} x = \operatorname{Re} z \\ y = \operatorname{Im} z \end{cases} \tag{2.5}$$

The complex numbers z with $y = 0$ are identified with the real numbers x, and thus the set of complex numbers includes the real numbers as a subset. When $x = 0$, in contrast, the complex numbers iy are called the pure imaginary numbers.

2.1.2 Algebraic Properties

By definition, two complex numbers are equal if and only if they have the same real parts and the same imaginary parts, that is,

$$x_1 + iy_1 = x_2 + iy_2 \quad \text{if and only if} \quad x_1 = x_2 \text{ and } y_1 = y_2 \qquad (2.6)$$

Algebraic properties of complex numbers are the same as for real numbers as listed below.

Commutative law of addition $z_1 + z_2 = z_2 + z_1$
Commutative law of multiplication $z_1 z_2 = z_2 z_1$
Associative law of addition $z_1 + (z_2 + z_3) = (z_1 + z_2) + z_3$
Associative law of multiplication $z_1(z_2 z_3) = (z_1 z_2)z_3$
Distributive law $z_1(z_2 + z_3) = z_1 z_2 + z_1 z_3$
Additive identity The complex number $0 = (0, 0)$ satisfies $z_1 + 0 = z_1$
Multiplicative identity The complex number $1 = (1, 0)$ satisfies $z_1 1 = z_1$
Additive inverse $z = -z_1$ if $z + z_1 = 0$
Multiplicative inverse $z = 1/z_1$ if $z z_1 = 1$

Operations with complex numbers can be performed as in the algebra of real numbers and by replacing i^2 by -1 when it occurs. Addition of two complex numbers $z_1 = x_1 + iy_1$ and $z_2 = x_2 + iy_2$ becomes

$$(x_1 + iy_1) + (x_2 + iy_2) = (x_1 + x_2) + i(y_1 + y_2) \qquad (2.7)$$

and subtraction is

$$(x_1 + iy_1) - (x_2 + iy_2) = (x_1 - x_2) + i(y_1 - y_2) \qquad (2.8)$$

Multiplication is given by

$$(x_1 + iy_1)(x_2 + iy_2) = x_1 x_2 + ix_1 y_2 + iy_1 x_2 + i^2 y_1 y_2$$
$$= (x_1 x_2 - y_1 y_2) + i(x_1 y_2 + x_2 y_1) \qquad (2.9)$$

where $i^2 = -1$ is used.

Division of z_1 by z_2 is obtained by multiplying the numerator and denominator by $x_2 - iy_2$ as

$$\frac{x_1 + iy_1}{x_2 + iy_2} = \frac{(x_1 + iy_1)(x_2 - iy_2)}{(x_2 + iy_2)(x_2 - iy_2)} = \frac{x_1 x_2 - ix_1 y_2 + iy_1 x_2 - i^2 y_1 y_2}{x_2^2 - i^2 y_2^2}$$
$$= \frac{x_1 x_2 + y_1 y_2}{x_2^2 + y_2^2} + i\frac{x_2 y_1 - x_1 y_2}{x_2^2 + y_2^2} \qquad (2.10)$$

where $i^2 = -1$ is used.

Example 2.1 Algebraic operations result in

$$\frac{5+10i}{3-4i} + \frac{25}{4+3i} = \frac{5+10i}{3-4i}\frac{3+4i}{3+4i} + \frac{25}{4+3i}\frac{4-3i}{4-3i}$$

$$= \frac{15+20i+30i+40i^2}{9-16i^2} + \frac{100-75i}{16-9i^2}$$

$$= \frac{-25+50i}{25} + \frac{100-75i}{25} = 3-i$$

which is expressed in standard form.

2.1.3 Complex Plane

Let us consider the geometric representation of complex numbers, which is of great practical importance. As a real number x is often represented by a point on an x line, it is natural to associate a complex number $z = x + iy$ with a point or a vector in an xy plane, as shown in Fig. 2.2.

The x and y axes are referred to as the real axis and imaginary axis, respectively, and the xy plane is referred to as the complex plane or the z plane. A complex number $z = x + iy$ is plotted as the point with coordinates (x, y). Each complex number corresponds to one and only one point in the complex plane, and, conversely, each point in the plane corresponds to one and only one complex number.

The absolute value or modulus of a complex number $z = x + iy$, denoted by $|z|$, is defined by

$$|z| = \sqrt{x^2 + y^2} \tag{2.11}$$

which is the length of the line segment from the origin to z, as shown in Fig. 2.2. If z is real, the modulus is simply the absolute value of z. The distance between two

Fig. 2.2 Complex plane representing complex numbers z and $\bar{z}$

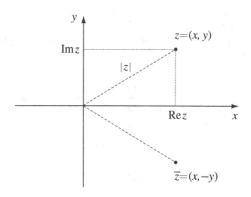

points $z_1 = x_1 + iy_1$ and $z_2 = x_2 + iy_2$ is given by

$$|z_1 - z_2| = \sqrt{(x_1 - x_2)^2 + (y_1 - y_2)^2} \qquad (2.12)$$

which is the length of the vector representing $z_1 - z_2$.

Example 2.2 The circle of radius R with its center at z_0 can be expressed as

$$|z - z_0| = R$$

For instance, the equation $|z - 1 + 2i| = 3$ represents the circle of radius $R = 3$ with its center at $1 - 2i$.

The complex conjugate of a complex number $z = x + iy$, denoted by $\bar{z}$, is defined by

$$\bar{z} = x - iy \qquad (2.13)$$

Geometrically, $\bar{z}$ is the point $(x, -y)$, obtained by reflecting $z = (x, y)$ in the real axis as shown in Fig. 2.2.

If $z_1 = x_1 + iy_1$ and $z_2 = x_2 + iy_2$, then

$$\overline{z_1 + z_2} = (x_1 + x_2) - i(y_1 + y_2) = (x_1 - iy_1) + (x_2 - iy_2) \qquad (2.14)$$

and thus the conjugate of the sum is the sum of the conjugates, that is,

$$\overline{z_1 + z_2} = \bar{z_1} + \bar{z_2} \qquad (2.15)$$

Similarly, the following identities hold:

$$\overline{z_1 - z_2} = \bar{z_1} - \bar{z_2} \qquad (2.16)$$

$$\overline{z_1 z_2} = \bar{z_1}\, \bar{z_2} \qquad (2.17)$$

$$\overline{\left(\frac{z_1}{z_2}\right)} = \frac{\bar{z_1}}{\bar{z_2}} \qquad (2.18)$$

By addition and subtraction, $z + \bar{z} = 2x$ and $z - \bar{z} = 2yi$, and it follows that

$$\begin{cases} \operatorname{Re} z = \dfrac{z + \bar{z}}{2} \\[2mm] \operatorname{Im} z = \dfrac{z - \bar{z}}{2i} \end{cases} \qquad (2.19)$$

By multiplication, $z\bar{z} = x^2 + y^2$, and the modulus of z is given by

$$|z| = \sqrt{z\bar{z}} \qquad (2.20)$$

The inverse of z is then given by

$$z^{-1} = \frac{1}{z} = \frac{\bar{z}}{z\bar{z}} = \frac{\bar{z}}{|z|^2} \tag{2.21}$$

provided $z \neq 0$.

Example 2.3 The circle of radius 1 with its center at $z = 1$, $x^2 + y^2 - 2x = 0$, can be expressed in terms of conjugate coordinates as

$$z\bar{z} - z - \bar{z} = 0$$

where the identities $z\bar{z} = x^2 + y^2$ and $x = (z + \bar{z})/2$ are used.

Example 2.4 The general equation for a circle or line in the xy plane is given by

$$a(x^2 + y^2) + bx + cy + d = 0$$

where a, b, c, and d are real constants, and $a \neq 0$ for a circle and $a = 0$ for a line. Using the aforementioned identities, it can be rewritten as

$$az\bar{z} + b\frac{z + \bar{z}}{2} + c\frac{z - \bar{z}}{2i} + d = 0$$

or, writing $a = \alpha$, $b/2 + c/(2i) = \beta$, and $d = \gamma$, it follows that

$$\alpha z\bar{z} + \beta z + \overline{\beta z} + \gamma = 0$$

which is the general equation for a circle or line in terms of conjugate coordinates.

2.1.4 Polar Form of Complex Numbers

A complex number z can also be expressed in terms of polar coordinates r and θ. The positive number r is the length of the vector representing z and θ is the angle that z (as a radius vector) makes with the positive real axis, as shown in Fig. 2.3.

Fig. 2.3 Polar representation of a complex number z

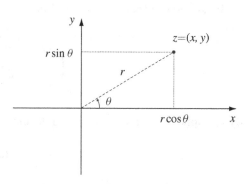

Using the identities

$$\begin{cases} x = r \cos \theta \\ y = r \sin \theta \end{cases} \qquad (2.22)$$

z can be expressed in polar form as

$$z = r(\cos \theta + i \sin \theta) \qquad (2.23)$$

where r is the modulus of z

$$r = |z| = \sqrt{x^2 + y^2} \qquad (2.24)$$

and θ is called the argument of z, denoted by arg z, such that

$$\tan \theta = \frac{y}{x} \qquad (2.25)$$

For a given $z \neq 0$, θ is determined only up to integer multiples of 2π and has any one of an infinite number of real values, as shown in Fig. 2.4, because of the periodic nature of cosine and sine with period of 2π.

To specify a unique value of arg z, the principal value Arg z is defined such that

$$-\pi < \text{Arg}\, z \leq \pi \qquad (2.26)$$

For a positive real number $z = x$, Arg $z = 0$, and for a negative real number, Arg $z = \pi$. The argument is given by

$$\arg z = \text{Arg}\, z + 2n\pi \qquad (2.27)$$

where n is an integer.

Example 2.5 For $z = 1 + i$, its polar form is

$$z = \sqrt{2}\,(\cos(\pi/4) + i \sin(\pi/4))$$

Fig. 2.4 Argument of
z given by Arg $z + 2n\pi$

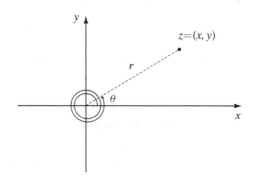

and $|z| = \sqrt{2}$, $\arg z = \pi/4 + 2n\pi$, and $\text{Arg } z = \pi/4$. For $z = -1 - i$, its polar form is

$$z = \sqrt{2}\left(\cos(-3\pi/4) + i\sin(-3\pi/4)\right)$$

and $|z| = \sqrt{2}$, $\arg z = -3\pi/4 + 2n\pi$, and $\text{Arg } z = -3\pi/4$.

The inverse of Eq. 2.25, given by

$$\theta = \arctan\frac{y}{x} \tag{2.28}$$

is not always true, and should be used with caution. Since $\tan\theta$ has period π, the arguments of z and $-z$ have the same tangent. The quadrant where z lies must be identified for a proper value.

Example 2.6 For $z = 1 + i$, it follows that

$$\theta = \arctan\frac{1}{1} = \arctan 1 = \pi/4$$

which is correct. On the other hand, for $z = -1 - i$, it follows that

$$\theta = \arctan\frac{-1}{-1} = \arctan 1 = \pi/4$$

which is not correct. However, by knowing z lies in the third quadrant and by using $\cos\theta = \sin\theta = -1/\sqrt{2}$, θ is obtained as $-3\pi/4$.

2.1.5 Exponential Form of Complex Numbers

Using the infinite series[1] of expansion $e^t = 1 + t + t^2/2! + t^3/3! + \cdots$ of elementary calculus with $t = i\theta$ yields

$$e^{i\theta} = \sum_{n=0}^{\infty}(-1)^n\frac{\theta^{2n}}{(2n)!} + i\sum_{n=0}^{\infty}(-1)^n\frac{\theta^{2n+1}}{(2n+1)!} \tag{2.29}$$

The first infinite series is $\cos\theta$ and the second is $\sin\theta$, and it follows that

$$e^{i\theta} = \cos\theta + i\sin\theta \tag{2.30}$$

which is known as Euler's formula. Applying this formula to a nonzero complex number z in polar form results in

$$z = re^{i\theta} \tag{2.31}$$

which is the exponential form of z.

---

[1] Series representations of complex variables are discussed in Chap. 5.

Example 2.7 For $z = 1 + i$ and $z = -1 - i$, the exponential forms are $z = \sqrt{2}e^{(\pi/4)i}$ and $z = \sqrt{2}e^{-(3\pi/4)i}$, respectively.

2.1.5.1 Multiplication in Exponential Form

Operations in exponential form give us a geometrical understanding of multiplication. Let $z_1 = r_1e^{i\theta_1}$ and $z_2 = r_2e^{i\theta_2}$, as shown in Fig. 2.5; then it follows that

$$z_1 z_2 = r_1 r_2 e^{i(\theta_1 + \theta_2)} \tag{2.32}$$

Taking absolute values on both sides reveals that the absolute value of a product is equal to the product of the absolute values of the factors:

$$|z_1 z_2| = |z_1||z_2| \tag{2.33}$$

Taking arguments on both sides reveals that the argument of a product is equal to the sum of the arguments of the factors:

$$\arg(z_1 z_2) = \arg z_1 + \arg z_2 \tag{2.34}$$

Figure 2.5 shows these geometrical relations.

2.1.5.2 Division in Exponential Form

Operations in exponential form give us a geometrical understanding of division, given by

$$\frac{z_1}{z_2} = \frac{r_1}{r_2} e^{i(\theta_1 - \theta_2)} \tag{2.35}$$

Fig. 2.5 Arguments of z_1, z_2, and $z_1 z_2$

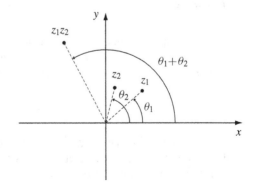

Taking absolute values on both sides reveals that the absolute value of a quotient is equal to the quotient of the absolute values of the factors:

$$\left| \frac{z_1}{z_2} \right| = \frac{|z_1|}{|z_2|} \tag{2.36}$$

Taking arguments on both sides reveals that the argument of a quotient is equal to the difference between the arguments of the factors:

$$\arg \frac{z_1}{z_2} = \arg z_1 - \arg z_2 \tag{2.37}$$

It should be noted that these properties are not, in general, valid when arg is replaced by Arg.

Example 2.8 Let us consider $z_1 = -1$ and $z_2 = i$ and specify $\arg z_1 = -\pi$ and $\arg z_2 = \pi/2$, then $\arg(z_1 z_2) = \arg(-i) = -\pi/2 = \arg z_1 + \arg z_2$. However, $\mathrm{Arg}\, z_1 = \pi$ and $\mathrm{Arg}\, z_2 = \pi/2$; thus, $\mathrm{Arg}\, z_1 + \mathrm{Arg}\, z_2 = 3\pi/2$, which is not equal to $\mathrm{Arg}(z_1 z_2) = \mathrm{Arg}(-i) = -\pi/2$.

Example 2.9 Let us consider a generalization of multiplication.

$$z_1 z_2 \ldots z_n = r_1 r_2 \ldots r_n e^{i(\theta_1 + \theta_2 + \cdots \theta_n)}$$

and if $z_1 = z_2 = \cdots = z_n = z$, then

$$z^n = r^n (\cos\theta + i\sin\theta)^n = r^n e^{in\theta} = r^n(\cos n\theta + i\sin n\theta)$$

The identity

$$(\cos\theta + i\sin\theta)^n = \cos n\theta + i\sin n\theta \tag{2.38}$$

is known as De Moivre's theorem.

2.1.6 Roots

Consider a complex number z_0 that satisfies $z^n = z_0$, where n is a positive integer. Then to a given $z_0 \neq 0$ there corresponds n distinct values of z, each of which is called an nth root of z_0. A set of n different roots is denoted by

$$z = z_0^{1/n} \tag{2.39}$$

which is n-valued. The symbol $z_0^{1/n}$ denotes n different roots. If z_0 is a positive real number r, $r^{1/n}$ denotes a set of nth roots, which must be distinguished from $\sqrt[n]{r}$ which is for a single-valued positive root.

Let $z_0 = r(\cos\theta + i\sin\theta)$ and $z = R(\cos\phi + i\sin\phi)$; then by De Moivre's theorem, $z^n = z_0$ is rewritten as

$$R^n(\cos n\phi + i\sin n\phi) = r(\cos\theta + i\sin\theta) \tag{2.40}$$

Equating the absolute values on both sides yields

$$R = \sqrt[n]{r} \tag{2.41}$$

where $\sqrt[n]{r}$ denotes the positive nth root of r. Equating the arguments on both sides yields

$$\phi = \frac{\theta + 2k\pi}{n} \tag{2.42}$$

where the periodic nature of cosine and sine with period of 2π is considered and $k = 0, 1, 2, \ldots, n-1$.

The n distinct values of the nth roots of $z_0 \neq 0$ are obtained as

$$z_0^{1/n} = \sqrt[n]{r}\left(\cos\frac{\theta + 2k\pi}{n} + i\sin\frac{\theta + 2k\pi}{n}\right) \tag{2.43}$$

which lie on a circle of radius $\sqrt[n]{r}$ with its center at the origin and constitute the vertices of a polygon of n sides. In particular, the root with $k = 0$ and $\theta = \mathrm{Arg}\,z_0$ is called the principal nth root of z_0.

When $z_0 = 1$, it follows that $r = 1$ and $\theta = 0$, and consequently, the nth roots are given by

$$1^{1/n} = \cos\frac{2k\pi}{n} + i\sin\frac{2k\pi}{n} \tag{2.44}$$

where $k = 0, 1, 2, \ldots, n-1$ and no further roots exist with other values of k. These n values are called the nth roots of unity, which lie on the circle of radius 1 with its center at the origin, called the unit circle.

Let ω denote the value corresponding to $k = 1$ in Eq. 2.44

$$\omega = \cos\frac{2\pi}{n} + i\sin\frac{2\pi}{n} \tag{2.45}$$

which is called the primitive nth root of unity. Then, according to De Moivre's theorem, it follows that

$$\omega^k = \cos\frac{2k\pi}{n} + i\sin\frac{2k\pi}{n} \tag{2.46}$$

Hence, the n values of $1^{1/n}$ are given by

$$1, \omega, \omega^2, \ldots, \omega^{n-1}$$

where 1 is the principal nth root of unity.

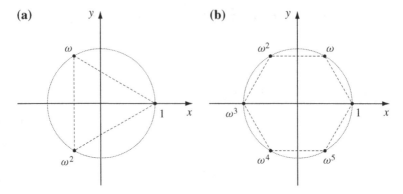

Fig. 2.6 Roots of unity. **a** Three cube roots. **b** Six 6th roots

Figure 2.6 shows the roots of unity for the cases of $n = 3$ and 6. The three cube roots of unity are on the vertices of an equilateral triangle and the six 6th roots of unity are on the vertices of a regular hexagon.

This property can be used to find the nth roots of any nonzero complex number z_0. In general, if α is any particular nth root of z_0 (not necessarily the principal root), then the n distinct values of $z_0^{1/n}$ are obtained as

$$\alpha, \alpha\omega, \alpha\omega^2, \ldots, \alpha\omega^{n-1}$$

since multiplication of α by ω^k corresponds to increasing the argument of α by $2k\pi/n$.

Example 2.10 Let us find all square roots of $-1 = 1(\cos \pi + i \sin \pi)$. From Eq. 2.43, it follows that

$$(-1)^{1/2} = \sqrt{1}\left(\cos \frac{\pi + 2k\pi}{2} + i \sin \frac{\pi + 2k\pi}{2}\right)$$

where $k = 0$, 1. The principal square root is i and the other root is $-$i. From Eq. 2.45, $\omega = -1$. Hence, if α is a particular square root, $-\alpha$ is the other root.

Example 2.11 Let us find all cube roots of $8 = 8(\cos 0 + i \sin 0)$. From Eq. 2.43, it follows that

$$8^{1/3} = \sqrt[3]{8}\left(\cos \frac{2k\pi}{3} + i \sin \frac{2k\pi}{3}\right)$$

where $k = 0$, 1, 2. The principal cube root is 2 and the other two roots are $-1 + \sqrt{3}i$ and $-1 - \sqrt{3}i$, as shown in Fig. 2.7. It is confirmed that three cube roots are on the vertices of an equilateral triangle, the arguments of which are different from each other by $2\pi/3$.

Fig. 2.7 The point $z = 8$ and three cube roots: 2, $-1 + \sqrt{3}i$, and $-1 - \sqrt{3}i$

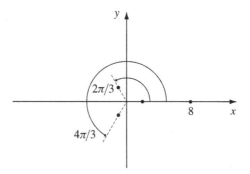

Example 2.12 Let us find all cube roots of $8i = 8(\cos \pi/2 + i \sin \pi/2)$. From Eq. 2.43, it follows that

$$(8i)^{1/3} = \sqrt[3]{8} \left(\cos \frac{\pi/2 + 2k\pi}{3} + i \sin \frac{\pi/2 + 2k\pi}{3} \right)$$

where $k = 0, 1, 2$. The principal cube root is $\sqrt{3} + i$ and the other two roots are $-\sqrt{3} + i$ and $-2i$, as shown in Fig. 2.8.

Knowing that a particular cube root of $8i$ is $\alpha = -2i$ and

$$\omega = \left(\cos \frac{2\pi}{3} + i \sin \frac{2\pi}{3} \right) = -\frac{1}{2} + \frac{\sqrt{3}}{2}i$$

it follows from Eq. 2.46 that the other two roots are

$$\alpha\omega = -2i \left(-\frac{1}{2} + \frac{\sqrt{3}}{2}i \right) = \sqrt{3} + i$$

Fig. 2.8 The point $z = 8i$ and three cube roots: $\sqrt{3} + i$, $-\sqrt{3} + i$, and $-2i$

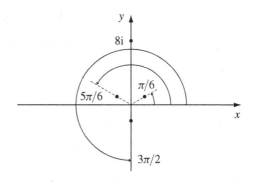

and

$$\alpha\omega^2 = -2i\left(-\frac{1}{2} + \frac{\sqrt{3}}{2}i\right)^2 = -\sqrt{3} + i$$

which, of course, are consistent with the solutions from Eq. 2.43.

2.2 Functions of a Complex Variable

Functions of a complex variable can be defined in a similar way to functions of a real variable. However, it should be noted that they operate on a complex variable rather than a real variable, and consequently exhibit interesting properties not shared by their real counterparts.

2.2.1 Definition

A function f defined on S, a set of complex numbers, is a rule that assigns to each value z belonging to S a complex number w. The number w is called the value of f at z, and this correspondence is denoted by

$$w = f(z) \tag{2.47}$$

where z is considered as a complex variable in the set S, called the domain of definition of f (or briefly the domain of f). The set of all values of a function f is called the range of f.

Suppose that $w = u + iv$ is the value of a function f at $z = x + iy$:

$$w = u + iv = f(z) = f(x + iy) \tag{2.48}$$

This relation implies that each of the real values u and v depends on the real variables x and y, which gives

$$\begin{cases} u = u(x, y) \\ v = v(x, y) \end{cases} \tag{2.49}$$

where $u(x, y)$ and $v(x, y)$ are the real functions of the real variables x and y.

Example 2.13 If $f(z) = z^2$, then

$$f(z) = (x + iy)^2 = x^2 - y^2 + 2xyi$$

and thus

$$\begin{cases} u(x, y) = x^2 - y^2 \\ v(x, y) = 2xy \end{cases}$$

The values of u and v depend on x and y.

It should be noted that a function $f(z)$, depending on the complex variable z, is equivalent to a pair of real functions $u(x, y)$ and $v(x, y)$, each depending on the two real variables x and y. In Example 2.13, a single complex function $f = z^2$ conveys two kinds of real functions, $u = x^2 - y^2$ and $v = 2xy$. This implies that a function of a complex variable contains two kinds of information (u and v), which could satisfy our needs addressed in Solution Strategy.

2.2.2 Elementary Functions

Many of the elementary functions appearing in real-variable calculus have natural complex extensions. The functions relevant to the engineering problems considered in this book are reviewed.

2.2.2.1 Polynomial Function

The polynomial function is defined by

$$w = \alpha_n z^n + \alpha_{n-1} z^{n-1} + \cdots + \alpha_1 z + \alpha_0 \tag{2.50}$$

where $\alpha_n \neq 0, \alpha_{n-1}, \ldots, \alpha_0$ are the complex constants and a positive integer n is the degree of the polynomial function. The domain of definition is the entire z plane.

2.2.2.2 Rational Function

The rational function is defined by

$$w = \frac{f(z)}{g(z)} \tag{2.51}$$

where $f(z)$ and $g(z)$ are the polynomial functions. The function $f(z)/g(z)$ is defined at each point z except where $g(z) = 0$.

2.2.2.3 Exponential Function

The exponential function is defined by

$$w = e^z = e^{x+iy} = e^x(\cos y + i \sin y) \tag{2.52}$$

where $e = 2.71828\ldots$ is the natural base of logarithms. The domain of definition is the entire z plane. Complex exponential functions have properties similar to those of real exponential functions, such as $e^{z_1+z_2} = e^{z_1}e^{z_2}$ and $e^{z_1-z_2} = e^{z_1}/e^{z_2}$.

2.2.2.4 Logarithmic Function

The inverse of the complex exponential function is the natural logarithmic function and is defined by

$$w = \ln z = \ln r + i\theta = \ln r + i(\theta + 2n\pi) \tag{2.53}$$

where $z = re^{i\theta} = re^{i(\theta+2n\pi)}$. The function $\ln z$ is defined at each nonzero point z. Since infinitely different values of $\arg z$ are obtained by successively encircling the origin $z = 0$, the complex logarithmic function is infinitely multi-valued. Each of the multiple functions is called a branch of the logarithmic function.

To keep the function single-valued, an artificial barrier that cannot be crossed is introduced. This barrier is called the branch cut, the direction of which is arbitrary. For instance, if the branch cut is set along the negative real half-axis (Fig. 2.9), $\arg z$ becomes single-valued and takes the principal value $-\pi < \text{Arg}\, z \leq \pi$. The value of $\ln z$ corresponding to the principal value is denoted by $\text{Ln}z$:

$$\text{Ln}z = \ln |z| + i\text{Arg}\, z \tag{2.54}$$

The values of $\arg z$ differ by integer multiples of 2π, and the other values of $\ln z$ is given by

$$\ln z = \text{Ln}z + 2n\pi i \tag{2.55}$$

where n is an integer. The point common to all branch cuts is called the branch point. The origin is a branch point of the logarithmic function.

Complex logarithmic functions have properties similar to those of real logarithmic functions, such as $\ln(z_1 z_2) = \ln z_1 + \ln z_2$ and $\ln(z_1/z_2) = \ln z_1 - \ln z_2$. It should be noted that these properties are not, in general, valid when $\ln$ is replaced by Ln.

Example 2.14 Let us consider $z_1 = z_2 = -1$ and specify $\ln z_1 = \pi i$ and $\ln z_2 = -\pi i$, then $\ln(z_1 z_2) = \ln 1 = 0$ and $\ln z_1 + \ln z_2 = 0$, thus the property $\ln(z_1 z_2) = \ln z_1 + \ln z_2$ is satisfied. In contrast, if the principal values are used, $\text{Ln}z_1 = \text{Ln}z_2 = \pi i$ and $\text{Ln}z_1 + \text{Ln}z_2 = 2\pi i$, which is not equal to $\text{Ln}(z_1 z_2) = \text{Ln}1 = 0$. Thus, the property $\text{Ln}(z_1 z_2) = \text{Ln}z_1 + \text{Ln}z_2$ is not always valid.

Fig. 2.9 Branch cut and branch point

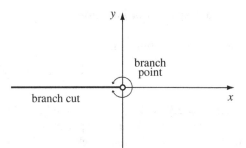

• Alternative Solution to Task 1-1

Using functions of a complex variable, Eq. 1.65 can be rewritten as

$$\Phi(z) = -\frac{q_u}{A}\operatorname{Re} z - \frac{q_s}{2\pi h}\operatorname{Re}[\ln(z - z_s)]$$
$$= \operatorname{Re}\left[-\frac{q_u}{A}z - \frac{q_s}{2\pi h}\ln(z - z_s)\right] \tag{2.56}$$

Substituting $q_u/A = 1$, $q_s/h = 6$, and $z_s = -1$ yields

$$\Phi(z) = \operatorname{Re}\left[-z - \frac{3}{\pi}\ln(z + 1)\right]$$

with which the equipotential lines are obtained as shown in Fig. 1.5.

2.3 Complex Differentiation

As is the case with functions of a real variable, the concepts of limit, continuity, and differentiability are important for functions of a complex variable. In developing a theory of differentiation for complex functions, the Cauchy–Riemann equations are introduced, which play a substantial role in complex analysis.

2.3.1 Limit and Continuity

A function $f(z)$ is said to have the limit w_0 at a point $z = z_0$ if

$$\lim_{z \to z_0} f(z) = w_0 \tag{2.57}$$

The point $w = f(z)$ can be made arbitrarily close to w_0 if the point z is chosen close enough to z_0 but distinct from it. In precise terms, for any positive number ε, some positive number δ can be found such that

$$|f(z) - w_0| < \varepsilon \quad \text{whenever} \quad 0 < |z - z_0| < \delta \tag{2.58}$$

As shown in Fig. 2.10, for each ε neighborhood[2] of w_0, there is a deleted δ neighborhood[3] of z_0 such that every point z in it has an image w lying in the ε neighborhood.

---

[2] An ε neighborhood of a point w_0 is the set of all points w such that $|w - w_0| < \varepsilon$ where ε is any given positive number.
[3] A deleted δ neighborhood of z_0 is a neighborhood of z_0 in which the point z_0 is omitted: $0 < |z - z_0| < \delta$.

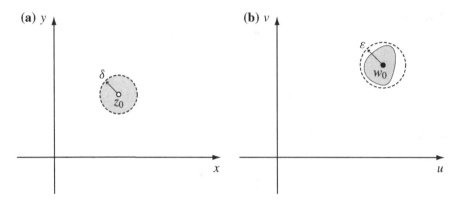

Fig. 2.10 Limit w_0. **a** Deleted δ neighborhood in the z plane. **b** ε neighborhood in the w plane

Because Eq. 2.58 applies to all points in the deleted neighborhood, the symbol $z \rightarrow z_0$ in Eq. 2.57 implies that z may approach z_0 from any direction in the complex plane.

A function $f(z)$ is said to be continuous at a point $z = z_0$ if

$$\lim_{z \to z_0} f(z) = f(z_0) \tag{2.59}$$

provided this limit and $f(z_0)$ exist. With the concept of limit, Eq. 2.59 can be rephrased that for any positive number ε, some positive number δ can be found such that

$$|f(z) - f(z_0)| < \varepsilon \quad \text{whenever} \quad |z - z_0| < \delta \tag{2.60}$$

If $f(z)$ is continuous at each point in a domain, $f(z)$ is said to be continuous in this domain.

With $z_0 = x_0 + iy_0$ and $f(z) = u + iv$, the real and imaginary parts of Eq. 2.59 can be written as

$$\begin{cases} \lim\limits_{\substack{x \to x_0 \\ y \to y_0}} u(x, y) = u(x_0, y_0) \\ \\ \lim\limits_{\substack{x \to x_0 \\ y \to y_0}} v(x, y) = v(x_0, y_0) \end{cases} \tag{2.61}$$

which states that the real functions u and v are continuous at (x_0, y_0). A function $f(z)$ is continuous if and only if its real and imaginary parts, u and v, are continuous.

Example 2.15 Let us consider $\lim_{z \to 0}(\bar{z}/z)$. If the limit is to exist, it must be independent of the approaching path of z to 0. Let $z \rightarrow 0$ along the x axis ($y = 0$), then it follows

$$\lim_{z \to 0} \frac{\bar{z}}{z} = \lim_{x \to 0} \frac{x}{x} = 1$$

On the other hand, if $z \rightarrow 0$ along the y axis ($x = 0$), then it follows that

$$\lim_{z \to 0} \frac{\overline{z}}{z} = \lim_{y \to 0} \frac{-iy}{iy} = -1$$

which is different from the previous value, and thus, the limit does not exist.

Example 2.16 Let us examine the continuity of $\overline{z} = x - iy$. Since $u(x, y) = x$ and $v(x, y) = -y$ are continuous at each point (x, y), $\overline{z}$ is continuous everywhere in the complex plane.

2.3.2 Differentiability

The derivative of a function $f(z)$ at a point z_0, denoted by $f'(z_0)$, is defined by

$$f'(z_0) = \lim_{z \to z_0} \frac{f(z) - f(z_0)}{z - z_0} \tag{2.62}$$

provided this limit exists. The function $f(z)$ is said to be differentiable at z_0 when its derivative at z_0 exists. As addressed in the definition of the limit, z may approach z_0 from any direction in the complex plane. Hence, differentiability at z_0 implies that the quotient in Eq. 2.62 always approaches a certain value and all these values are equal.

With the increment $\Delta z = z - z_0$, Eq. 2.62 takes the form

$$f'(z_0) = \lim_{\Delta z \to 0} \frac{f(z_0 + \Delta z) - f(z_0)}{\Delta z} \tag{2.63}$$

Let $w = f(z)$ and $\Delta w = f(z_0 + \Delta z) - f(z_0)$, then Eq. 2.63 reduces to

$$f'(z_0) = \frac{dw}{dz} = \lim_{\Delta z \to 0} \frac{\Delta w}{\Delta z} \tag{2.64}$$

where $dw = f'(z_0)dz$ is the differential of w and dz is the differential of z. It should be noted that dz and dw are not the limits of Δz and Δw as $\Delta z \to 0$; these limits are zero whereas dz and dw are not necessarily zero. The differential dw is a dependent variable determined through $dw = f'(z_0)dz$ with the independent variable dz for a given z_0.

The differentiation rules for functions of a complex variable are the same as in real-variable calculus. For any differentiable functions f and g, the following rules are valid.

Linearity rule If $w = c_1 f + c_2 g$, where c_1 and c_2 are constants, then

$$w' = (c_1 f + c_2 g)' = c_1 f' + c_2 g' \tag{2.65}$$

Product rule If $w = fg$, then

$$w' = (fg)' = f'g + fg' \tag{2.66}$$

Quotient rule If $w = f/g$, where $g \neq 0$, then

$$w' = \left(\frac{f}{g}\right)' = \frac{f'g - fg'}{g^2} \tag{2.67}$$

Chain rule If $w = f(\chi)$ and $\chi = g(z)$, then

$$\frac{dw}{dz} = \frac{dw}{d\chi}\frac{d\chi}{dz} \tag{2.68}$$

Example 2.17 Let us prove the chain rule. Consider the increments given by

$$\begin{cases} \Delta w = f(\chi + \Delta\chi) - f(\chi) \\ \Delta\chi = g(z + \Delta z) - g(z) \end{cases}$$

where $\Delta w \to 0$ and $\Delta\chi \to 0$ as $\Delta z \to 0$. If $\Delta\chi \neq 0$, let us define ε as

$$\varepsilon = \frac{\Delta w}{\Delta\chi} - \frac{dw}{d\chi} \tag{2.69}$$

so that $\varepsilon \to 0$ as $\Delta\chi \to 0$. Rearranging Eq. 2.69 for Δw gives

$$\Delta w = \frac{dw}{d\chi}\Delta\chi + \varepsilon\Delta\chi \tag{2.70}$$

If $\Delta\chi = 0$, according to the definition of Δw, $\Delta w = 0$, and ε is set as zero. Hence, for any $\Delta\chi$ ($\Delta\chi \neq 0$ or $\Delta\chi = 0$), Eq. 2.70 holds.

Dividing Eq. 2.70 by $\Delta z \neq 0$ and in the limit of $\Delta z \to 0$

$$\lim_{\Delta z \to 0}\frac{\Delta w}{\Delta z} = \frac{dw}{dz} = \lim_{\Delta z \to 0}\left(\frac{dw}{d\chi}\frac{\Delta\chi}{\Delta z} + \varepsilon\frac{\Delta\chi}{\Delta z}\right)$$

$$= \frac{dw}{d\chi}\lim_{\Delta z \to 0}\frac{\Delta\chi}{\Delta z} + \lim_{\Delta z \to 0}\varepsilon\lim_{\Delta z \to 0}\frac{\Delta\chi}{\Delta z}$$

$$= \frac{dw}{d\chi}\frac{d\chi}{dz} + \lim_{\Delta z \to 0}\varepsilon\frac{d\chi}{dz}$$

$$= \frac{dw}{d\chi}\frac{d\chi}{dz}$$

where $\varepsilon \to 0$ in the limit of $\Delta z \to 0$ is used, and the chain rule is proved.

Example 2.18 Let us find the derivative of $w = z^2$. The increment Δw is

$$\Delta w = (z + \Delta z)^2 - z^2 = z^2 + 2z\Delta z + \Delta z^2 - z^2$$
$$= 2z\Delta z + \Delta z^2$$

From Eq. 2.64, it follows that

$$w' = \lim_{\Delta z \to 0} \frac{2z\Delta z + \Delta z^2}{\Delta z} = 2z$$

which is identical to the differentiation in real-variable calculus. The power rule of differentiation

$$(z^n)' = nz^{n-1} \tag{2.71}$$

generally holds for complex functions.

Example 2.19 Let us find the derivative of $w = 1/z$. The increment Δw is

$$\Delta w = \frac{1}{z + \Delta z} - \frac{1}{z} = \frac{z - z - \Delta z}{(z + \Delta z)z}$$
$$= -\frac{\Delta z}{(z + \Delta z)z}$$

From Eq. 2.64, it follows that

$$w' = -\lim_{\Delta z \to 0} \frac{\Delta z}{\Delta z(z + \Delta z)z} = -\frac{1}{z^2}$$

which is identical to the differentiation in real-variable calculus. The power rule of differentiation

$$(z^{-n})' = -nz^{-n-1} \tag{2.72}$$

generally holds for complex functions.

2.3.3 Analytic Functions

A function $f(z)$ is said to be analytic at a point z_0 when there exists a neighborhood $|z - z_0| < \delta$ for all points of which $f'(z)$ exists. That is to say, if $f(z)$ is analytic at z_0, $f(z)$ must be differentiable not only at z_0, but also at all points in some δ neighborhood of z_0. This concept is raised because differentiability of $f(z)$ merely at a single point z_0 is of no practical interest.

When the derivative $f'(z)$ exists at all points z in a domain, a function $f(z)$ is said to be analytic in this domain and is referred to as an analytic function. In particular, if $f(z)$ is analytic at all points in the entire complex plane, then $f(z)$ is referred to as an entire function.

Example 2.20 Let us examine the analyticity of $w = z^2$. In Example 2.18, the derivative is obtained as $w' = 2z$ and it is obvious that z^2 is analytic at all points in the entire plane. The function z^2 is entire.

Example 2.21 Let us examine the analyticity of $w = |z|^2$. The increment Δw is

$$\Delta w = |z + \Delta z|^2 - |z|^2 = (z + \Delta z)(\bar{z} + \overline{\Delta z}) - z\bar{z}$$
$$= z\overline{\Delta z} + \bar{z}\Delta z + \overline{\Delta z}\Delta z$$

From Eq. 2.64, it follows that

$$w' = \lim_{\Delta z \to 0} \frac{z\overline{\Delta z} + \bar{z}\Delta z + \overline{\Delta z}\Delta z}{\Delta z} = \bar{z} + \lim_{\Delta z \to 0} \left(z\frac{\overline{\Delta z}}{\Delta z} + \overline{\Delta z} \right)$$

If the limit exists, it can be found by letting $\Delta z = (\Delta x, \Delta y)$ approach 0 in any manner. Setting $\Delta y = 0$ and letting $\Delta z = (\Delta x, 0)$ approach 0 yields $\overline{\Delta z} = \Delta z$, which gives the value of derivative $w' = \bar{z} + z$. Similarly, setting $\Delta x = 0$ and letting $\Delta z = (0, \Delta y)$ approach 0 yields $\overline{\Delta z} = -\Delta z$, which gives the value of derivative $w' = \bar{z} - z$. Since limits are unique, $\bar{z} + z$ must be equal to $\bar{z} - z$, and it follows that $z = 0$ and $w' = 0$. The function $|z|^2$ has the derivative only at the point $z = 0$. It is not differentiable at any point in a neighborhood of $z = 0$, and thus, $|z|^2$ is not analytic.

Example 2.22 Let us examine the analyticity of $w = 1/z$. In Example 2.19, the derivative is obtained as $w' = -1/z^2$ and it is obvious that $1/z$ is analytic at all nonzero points in the complex plane.

If a function $f(z)$ is not analytic at a point z_0, but every neighborhood of z_0 contains at least one point where $f(z)$ is analytic, then the point z_0 is called the singular point or singularity of $f(z)$. For instance, the function $1/z$ has a singular point at $z = 0$, since $1/z$ is analytic at all nonzero points in the complex plane. On the other hand, the function $|z|^2$ is nowhere analytic and has no singularity.

2.3.4 L'Hospital's Rule

If $f(z)$ is analytic, Eq. 2.62 can be rewritten as

$$\frac{f(z) - f(z_0)}{z - z_0} - f'(z) = \varepsilon \tag{2.73}$$

where $\varepsilon \to 0$ as $z \to z_0$. It follows that

$$f(z) = f(z_0) + f'(z_0)(z - z_0) + \varepsilon(z - z_0) \tag{2.74}$$

Now, let us consider analytic functions $f(z)$ and $g(z)$ in a domain containing z_0 and suppose that $f(z_0) = g(z_0) = 0$ and $g'(z_0) \neq 0$. Then, using Eq. 2.74 and the fact that $f(z_0) = g(z_0) = 0$, it follows that

$$\begin{cases} f(z) = f'(z_0)(z - z_0) + \varepsilon_1(z - z_0) \\ g(z) = g'(z_0)(z - z_0) + \varepsilon_2(z - z_0) \end{cases} \tag{2.75}$$

where $\varepsilon_1 \to 0$ and $\varepsilon_2 \to 0$ as $z \to z_0$. Then

$$\lim_{z \to z_0} \frac{f(z)}{g(z)} = \frac{f'(z_0)}{g'(z_0)} = \lim_{z \to z_0} \frac{f'(z)}{g'(z)} \tag{2.76}$$

which is known as L'Hospital's rule which often converts the quotient to a determinate form and allows the limit to be evaluated.

Example 2.23 Let us evaluate

$$\lim_{z \to i} \frac{f(z)}{g(z)} = \lim_{z \to i} \frac{z^7 + i}{z^3 + i}$$

Since $f(i) = g(i) = 0$ and $f(z)$ and $g(z)$ are analytic at $z = i$, L'Hospital's rule can be applied as

$$\lim_{z \to i} \frac{f'(z)}{g'(z)} = \frac{7i^6}{3i^2} = \frac{7}{3}i^4 = \frac{7}{3}$$

and the limit is obtained as $7/3$.

Example 2.24 In the case $f'(z_0) = g'(z_0) = 0$, L'Hospital's rule can be extended. Let us evaluate

$$\lim_{z \to 0} \frac{f(z)}{g(z)} = \lim_{z \to 0} \frac{1 - \cos z}{z^2}$$

Since $f(0) = g(0) = 0$ and $f(z)$ and $g(z)$ are analytic at $z = 0$, L'Hospital's rule can be applied as

$$\lim_{z \to 0} \frac{f'(z)}{g'(z)} = \lim_{z \to 0} \frac{\sin z}{2z}$$

where $\sin z$ and $2z$ are analytic and equal to 0 when $z = 0$. By applying L'Hospital's rule again, it follows that

$$\lim_{z \to 0} \frac{f''(z)}{g''(z)} = \frac{\cos 0}{2} = \frac{1}{2}$$

and the limit is obtained as $1/2$.

2.3.5 Cauchy–Riemann Equations in Cartesian Form

To perform complex analysis on any function, the analyticity of the function is essential. Let us assume the derivative $f'(z)$ at z exists as

$$f'(z) = \lim_{\Delta z \to 0} \frac{f(z + \Delta z) - f(z)}{\Delta z} \tag{2.77}$$

By the definition of the limit, Δz can approach zero along any path in a neighborhood of z. If the approaching path is set horizontally, $\Delta z = \Delta x$, Eq. 2.77 can be written as

$$
\begin{aligned}
f'(z) &= \lim_{\Delta x \to 0} \frac{f(z + \Delta x) - f(z)}{\Delta x} \\
&= \lim_{\Delta x \to 0} \frac{u(x + \Delta x, y) + iv(x + \Delta x, y) - u(x, y) - iv(x, y)}{\Delta x} \\
&= \lim_{\Delta x \to 0} \left[\frac{u(x + \Delta x, y) - u(x, y)}{\Delta x} + i \frac{v(x + \Delta x, y) - v(x, y)}{\Delta x} \right] \\
&= \frac{\partial u}{\partial x} + i \frac{\partial v}{\partial x} \tag{2.78}
\end{aligned}
$$

provided that the partial derivatives exist.

Similarly, if the approaching path is set vertically, $\Delta z = i\Delta y$, Eq. 2.77 can be written as

$$
\begin{aligned}
f'(z) &= \lim_{\Delta y \to 0} \frac{f(z + i\Delta y) - f(z)}{i\Delta y} \\
&= \lim_{\Delta y \to 0} \frac{u(x, y + \Delta y) + iv(x, y + \Delta y) - u(x, y) - iv(x, y)}{i\Delta y} \\
&= \lim_{\Delta y \to 0} \left[\frac{u(x, y + \Delta y) - u(x, y)}{i\Delta y} + i \frac{v(x, y + \Delta y) - v(x, y)}{i\Delta y} \right] \\
&= -i \frac{\partial u}{\partial y} + \frac{\partial v}{\partial y} \tag{2.79}
\end{aligned}
$$

where the identity $1/i = -i$ is used.

For the existence of the derivative $f'(z)$, the values of Eqs. 2.78 and 2.79 must be equal. By equating the real and imaginary parts on the right-hand sides of these equations, necessary conditions for the existence of $f'(z)$ can be derived as

$$
\begin{cases}
\dfrac{\partial u}{\partial x} = \dfrac{\partial v}{\partial y} \\[2mm]
\dfrac{\partial u}{\partial y} = -\dfrac{\partial v}{\partial x}
\end{cases} \tag{2.80}
$$

which are known as the Cauchy–Riemann equations.

These observations imply the necessity of the Cauchy–Riemann equations for a function to be analytic and can be summarized as follows.

Theorem 2.1 (Cauchy–Riemann equations: Necessity) *If $f(z) = u + iv$ is analytic in a domain, the first-order partial derivatives of u and v with respect to x and y exist and satisfy the Cauchy–Riemann equations at all points in this domain.*

Example 2.25 Let us consider the analytic function $f(z) = z^2$. Since

$$\begin{cases} u(x, y) = x^2 - y^2 \\ v(x, y) = 2xy \end{cases}$$

the partial derivatives are

$$\begin{cases} \partial u/\partial x = 2x \\ \partial u/\partial y = -2y \end{cases}$$

and

$$\begin{cases} \partial v/\partial y = 2x \\ -\partial v/\partial x = -2y \end{cases}$$

which satisfy the Cauchy–Riemann equations. The derivative is

$$f'(z) = \frac{\partial u}{\partial x} + i\frac{\partial v}{\partial x} = 2x + 2yi = 2(x + iy) = 2z$$

which is identical to the differentiation in real-variable calculus. As also noted in Example 2.18, the power rule of differentiation generally holds for complex functions.

Example 2.26 Let us consider the function $f(z) = |z|^2$. Since

$$\begin{cases} u(x, y) = x^2 + y^2 \\ v(x, y) = 0 \end{cases}$$

the partial derivatives are

$$\begin{cases} \partial u/\partial x = 2x \\ \partial u/\partial y = 2y \end{cases}$$

and

$$\begin{cases} \partial v/\partial y = 0 \\ -\partial v/\partial x = 0 \end{cases}$$

which do not satisfy the Cauchy–Riemann equations except at 0, and thus, $f(z) = |z|^2$ is not analytic. The derivative is

$$f'(z) = \frac{\partial u}{\partial x} + i\frac{\partial v}{\partial x} = 0$$

only at $z = 0$.

In real-variable analysis, the functions x^2 and $|x|^2$ are identical and have the same derivative of $2x$. The very different conclusions of Examples 2.25 and 2.26 exemplify an interesting feature of differentiability of functions of a complex variable.

It should be noted that differentiability of $f(z) = u + iv$ is not ensured by individual differentiability of u and v. Functions u and v need to be related through the Cauchy–Riemann equations. This is different from continuity of $f(z)$, which is equivalent to individual continuity of u and v.

Example 2.27 Let us consider real functions $u(x, y) = x$ and $v(x, y) = -y$, which are differentiable. The partial derivatives are

$$\begin{cases} \partial u/\partial x = 1 \\ \partial u/\partial y = 0 \end{cases}$$

and

$$\begin{cases} \partial v/\partial y = -1 \\ -\partial v/\partial x = 0 \end{cases}$$

which do not satisfy the Cauchy–Riemann equations, and $\bar{z} = x - iy$ is not analytic.

It should be noted that $\bar{z}$ is not differentiable anywhere even though its real and imaginary parts (u and v) are continuous (and therefore $\bar{z}$ is continuous as shown in Example 2.16) and differentiable.

Theorem 2.1 states that the Cauchy–Riemann equations are necessary for the existence of the derivative $f'(z)$ but does not cover the sufficiency. However, if the first partial derivatives in Eq. 2.80 are continuous in a domain, then it can be shown that the Cauchy–Riemann equations are sufficient conditions that $f(z)$ be analytic in this domain.

Theorem 2.2 (Cauchy–Riemann equations: Sufficiency) *If u and v have continuous first-order partial derivatives with respect to x and y that satisfy the Cauchy–Riemann equations in a domain, the function $f(z) = u + iv$ is analytic in this domain.*

Proof In view of the continuity of the first-order partial derivatives of u with respect to x and y, the increment Δu is given by

$$\Delta u = u(x + \Delta x, y + \Delta y) - u(x, y)$$

$$= u(x + \Delta x, y + \Delta y) - u(x, y + \Delta y) + u(x, y + \Delta y) - u(x, y)$$

$$= \left(\frac{\partial u}{\partial x} + \varepsilon_u \right) \Delta x + \left(\frac{\partial u}{\partial y} + \eta_u \right) \Delta y$$

$$= \frac{\partial u}{\partial x} \Delta x + \frac{\partial u}{\partial y} \Delta y + \varepsilon_u \Delta x + \eta_u \Delta y$$

where the identity Eq. 2.74 is used with $\varepsilon_u \to 0$ as $\Delta x \to 0$ and $\eta_u \to 0$ as $\Delta y \to 0$.

Similarly, in view of the continuity of the first-order partial derivatives of v with respect to x and y, the increment Δv is given by

$$\Delta v = v(x + \Delta x, y + \Delta y) - v(x, y)$$

$$= \frac{\partial v}{\partial x}\Delta x + \frac{\partial v}{\partial y}\Delta y + \varepsilon_v \Delta x + \eta_v \Delta y$$

where $\varepsilon_v \to 0$ as $\Delta x \to 0$ and $\eta_v \to 0$ as $\Delta y \to 0$.

The increment Δw is then given by

$$\Delta w = \Delta u + i\Delta v$$

$$= \left(\frac{\partial u}{\partial x} + i\frac{\partial v}{\partial x}\right)\Delta x + \left(\frac{\partial u}{\partial y} + i\frac{\partial v}{\partial y}\right)\Delta y + \varepsilon \Delta x + \eta \Delta y$$

where $\varepsilon = \varepsilon_u + i\varepsilon_v \to 0$ and $\eta = \eta_u + i\eta_v \to 0$ as $\Delta x \to 0$ and $\Delta y \to 0$.

Assuming that the Cauchy–Riemann equations are satisfied, $\partial u/\partial y$ is replaced by $-\partial v/\partial x$ and $\partial v/\partial y$ by $\partial u/\partial x$, and then

$$\Delta w = \left(\frac{\partial u}{\partial x} + i\frac{\partial v}{\partial x}\right)\Delta x + \left(\frac{-\partial v}{\partial x} + i\frac{\partial u}{\partial x}\right)\Delta y + \varepsilon \Delta x + \eta \Delta y$$

$$= \left(\frac{\partial u}{\partial x} + i\frac{\partial v}{\partial x}\right)(\Delta x + i\Delta y) + \varepsilon \Delta x + \eta \Delta y$$

$$= \left(\frac{\partial u}{\partial x} + i\frac{\partial v}{\partial x}\right)\Delta z + \varepsilon \Delta x + \eta \Delta y$$

Dividing by Δz and in the limit of $\Delta z \to 0$

$$\lim_{\Delta z \to 0} \frac{\Delta w}{\Delta z} = f'(z) = \frac{\partial u}{\partial x} + i\frac{\partial v}{\partial x}$$

which indicates that the derivative exists and the function $f(z)$ is analytic. □

Example 2.28 Let us consider the property of a function $f(z)$, when $f(z)$ is analytic in a domain and $|f(z)|$ is constant in this domain. It follows that

$$|f(z)|^2 = |u + iv|^2 = u^2 + v^2 = \text{constant}$$

Taking the partial derivatives with respect to x and y yields

$$\begin{cases} u(\partial u/\partial x) + v(\partial v/\partial x) = 0 \\ u(\partial u/\partial y) + v(\partial v/\partial y) = 0 \end{cases}$$

From the Cauchy–Riemann equations, it follows that

$$\begin{cases} u(\partial u/\partial x) - v(\partial u/\partial y) = 0 \\ u(\partial u/\partial y) + v(\partial u/\partial x) = 0 \end{cases}$$

which gives

$$\begin{cases} (u^2 + v^2)(\partial u/\partial x) = 0 \\ (u^2 + v^2)(\partial u/\partial y) = 0 \end{cases}$$

This implies that $u^2 + v^2 = 0$ or $\partial u/\partial x = \partial u/\partial y = 0$. If $u^2 + v^2 = 0$, then $u = v = 0$ and $f(z) = 0 = $ constant. If $u^2 + v^2 \neq 0$, from the Cauchy–Riemann equations, then also $\partial v/\partial x = \partial v/\partial y = 0$. Hence, u is constant and v is constant; consequently, $f(z)$ is constant. In either case, if $|f(z)|$ is constant, $f(z)$ is constant.

2.3.6 Cauchy–Riemann Equations in Polar Form

When complex functions are expressed in polar coordinates, the Cauchy–Riemann equations in polar form are of practical use. Let us assume the derivative $f'(z)$ at $z = re^{i\theta}$ exists, which is given by Eq. 2.77. By the definition of the limit, Δz can approach zero along any path in a neighborhood of z. If the approaching path is set along the ray θ, $\Delta z = \Delta r e^{i\theta}$, Eq. 2.77 can be written as

$$
\begin{aligned}
f'(z) &= \lim_{\Delta r \to 0} \frac{f((r + \Delta r)e^{i\theta}) - f(re^{i\theta})}{\Delta r e^{i\theta}} \\
&= \lim_{\Delta r \to 0} \frac{u(r + \Delta r, \theta) + iv(r + \Delta r, \theta) - u(r, \theta) - iv(r, \theta)}{\Delta r e^{i\theta}} \\
&= \lim_{\Delta r \to 0} \left[\frac{u(r + \Delta r, \theta) - u(r, \theta)}{\Delta r e^{i\theta}} + i\frac{v(r + \Delta r, \theta) - v(r, \theta)}{\Delta r e^{i\theta}} \right] \\
&= e^{-i\theta} \left(\frac{\partial u}{\partial r} + i\frac{\partial v}{\partial r} \right)
\end{aligned}
\tag{2.81}
$$

provided that the partial derivatives exist.

Similarly, if the approaching path is set along the circle r, $\Delta z = r(e^{i(\theta + \Delta\theta)} - e^{i\theta})$, Eq. 2.77 can be written as

$$
\begin{aligned}
f'(z) &= \lim_{\Delta\theta \to 0} \frac{f(re^{i(\theta + \Delta\theta)}) - f(re^{i\theta})}{r(e^{i(\theta + \Delta\theta)} - e^{i\theta})} \\
&= \lim_{\Delta\theta \to 0} \frac{u(r, \theta + \Delta\theta) + iv(r, \theta + \Delta\theta) - u(r, \theta) - iv(r, \theta)}{re^{i\theta}\Delta\theta} \frac{\Delta\theta}{e^{i\Delta\theta} - 1} \\
&= \lim_{\Delta\theta \to 0} \left[\frac{u(r, \theta + \Delta\theta) - u(r, \theta)}{re^{i\theta}\Delta\theta} + i\frac{v(r, \theta + \Delta\theta) - v(r, \theta)}{re^{i\theta}\Delta\theta} \right] \frac{\Delta\theta}{e^{i\Delta\theta} - 1} \\
&= e^{-i\theta} \left(\frac{1}{r}\frac{\partial u}{\partial\theta} + i\frac{1}{r}\frac{\partial v}{\partial\theta} \right)\frac{1}{i} = e^{-i\theta} \left(-i\frac{1}{r}\frac{\partial u}{\partial\theta} + \frac{1}{r}\frac{\partial v}{\partial\theta} \right)
\end{aligned}
\tag{2.82}
$$

where L'Hospital's rule is used to obtain $\lim_{\Delta\theta \to 0} \Delta\theta/(e^{i\Delta\theta} - 1) = 1/i$.

For the existence of the derivative $f'(z)$, the values of Eqs. 2.81 and 2.82 must be equal. Equating the real and imaginary parts on the right-hand sides of these equations yields

$$\begin{cases} \dfrac{\partial u}{\partial r} = \dfrac{1}{r}\dfrac{\partial v}{\partial \theta} \\[3mm] \dfrac{1}{r}\dfrac{\partial u}{\partial \theta} = -\dfrac{\partial v}{\partial r} \end{cases} \tag{2.83}$$

which are the Cauchy–Riemann equations in polar form.

Example 2.29 Let us consider the function $f(z) = \ln z = \ln r + i\theta$. Since

$$\begin{cases} u(r, \theta) = \ln r \\ v(r, \theta) = \theta \end{cases}$$

the partial derivatives are

$$\begin{cases} \partial u/\partial r = 1/r \\ (1/r)(\partial u/\partial \theta) = 0 \end{cases}$$

and

$$\begin{cases} (1/r)(\partial v/\partial \theta) = 1/r \\ -\partial v/\partial r = 0 \end{cases}$$

which satisfy the Cauchy–Riemann equations, and $f(z) = \ln z$ is analytic. The derivative is

$$f'(z) = e^{-i\theta}(\partial u/\partial r + i\partial v/\partial r) = e^{-i\theta}(1/r) = 1/(re^{i\theta}) = 1/z$$

which is identical to the differentiation in real-variable calculus.

Example 2.30 Let us consider the function $f(z) = 1/z = r^{-1}e^{-i\theta}$. Since

$$\begin{cases} u(r, \theta) = \cos\theta/r \\ v(r, \theta) = -\sin\theta/r \end{cases}$$

the partial derivatives are

$$\begin{cases} \partial u/\partial r = -\cos\theta/r^2 \\ (1/r)(\partial u/\partial \theta) = -\sin\theta/r^2 \end{cases}$$

and

$$\begin{cases} (1/r)(\partial v/\partial \theta) = -\cos\theta/r^2 \\ -\partial v/\partial r = -\sin\theta/r^2 \end{cases}$$

which satisfy the Cauchy–Riemann equations, and $f(z) = 1/z$ is analytic. The derivative is

$$f'(z) = e^{-i\theta} \left(\partial u/\partial r + i\partial v/\partial r \right) = e^{-i\theta} \left(-\cos\theta/r^2 + i\sin\theta/r^2 \right)$$

$$= -e^{-i\theta}e^{-i\theta}/r^2 = -1/(re^{i\theta})^2 = -1/z^2$$

which is identical to the differentiation in real-variable calculus.

2.4 Harmonic Functions

As discussed in Chap. 1, Laplace's equation occurs in many engineering problems and plays a central role in potential theory. Functions that fulfill Laplace's equation are called harmonic functions, which are discussed in this section. In particular, analyticity of complex functions and the existence of a harmonic conjugate are of great practical importance.

2.4.1 Analyticity of Elementary Functions

Analyticity of the elementary functions reviewed in Sect. 2.2.2 is examined.

2.4.1.1 Polynomial Function

The power rule of differentiation (identical to that in real-variable calculus) holds for functions of a complex variable, and thus

$$\frac{d(\alpha_n z^n + \alpha_{n-1}z^{n-1} + \cdots + \alpha_1 z + \alpha_0)}{dz}$$

$$= n\alpha_n z^{n-1} + (n-1)\alpha_{n-1}z^{n-2} + \cdots + \alpha_1 \tag{2.84}$$

and the polynomial function is analytic in the entire complex plane.

2.4.1.2 Rational Function

The quotient rule of differentiation (identical to that in real-variable calculus) holds for functions of a complex variable, and thus

$$\frac{d}{dz}\frac{f(z)}{g(z)} = \frac{f'(z)g(z) - f(z)g'(z)}{g(z)^2} \tag{2.85}$$

and the rational function is analytic except at the points where $g(z) = 0$.

2.4.1.3 Exponential Function

By definition (Eq. 2.52), $w = e^x(\cos y + i \sin y) = u + iv$, then

$$\begin{cases} \dfrac{\partial u}{\partial x} = e^x \cos y = \dfrac{\partial v}{\partial y} \\[3mm] \dfrac{\partial u}{\partial y} = -e^x \sin y = -\dfrac{\partial v}{\partial x} \end{cases} \tag{2.86}$$

indicating the Cauchy–Riemann equations are satisfied. Hence, the required derivative exists and is given by

$$\frac{d}{dz}e^z = \frac{\partial u}{\partial x} + i\frac{\partial v}{\partial x} = e^x \cos y + ie^x \sin y = e^z \tag{2.87}$$

which is identical to the result in real-variable calculus and the exponential function is analytic in the entire complex plane.

2.4.1.4 Logarithmic Function

Let $w = \ln z$, then $z = e^w$ and $dz/dw = e^w = z$, thus

$$\frac{d}{dz}\ln z = \frac{dw}{dz} = \frac{1}{dz/dw} = \frac{1}{z} \tag{2.88}$$

which is identical to the result in real-variable calculus and consistent with Example 2.29. The logarithmic function is analytic except at the branch point, $z = 0$, and on the branch cut, the negative real half-axis. The origin and each point on the negative real half-axis are the singular points of $\ln z$.

2.4.2 Laplace's Equation in Cartesian Form

If a function $f(z) = u + iv$ is analytic in a domain, the Cauchy–Riemann equations are satisfied in this domain. As is to be proved in Chap. 4 (Corollary 4.3), it is possible to assume that u and v have continuous second partial derivatives.

Differentiating both sides of the first Cauchy–Riemann equation in Cartesian form (Eq. 2.80) with respect to x and both sides of the second with respect to y yields

$$\begin{cases} \dfrac{\partial^2 u}{\partial x^2} = \dfrac{\partial^2 v}{\partial x \partial y} \\[3mm] \dfrac{\partial^2 u}{\partial y^2} = -\dfrac{\partial^2 v}{\partial x \partial y} \end{cases} \tag{2.89}$$

Adding these two equations gives Laplace's equation in terms of u:

$$\frac{\partial^2 u}{\partial x^2} + \frac{\partial^2 u}{\partial y^2} = 0 \tag{2.90}$$

Similarly, differentiating both sides of the first Cauchy–Riemann equation with respect to y and both sides of the second with respect to x yields

$$\begin{cases} \dfrac{\partial^2 u}{\partial x \partial y} = \dfrac{\partial^2 v}{\partial y^2} \\[3mm] \dfrac{\partial^2 u}{\partial x \partial y} = -\dfrac{\partial^2 v}{\partial x^2} \end{cases} \tag{2.91}$$

Subtracting the second equation from the first gives Laplace's equation in terms of v:

$$\frac{\partial^2 v}{\partial x^2} + \frac{\partial^2 v}{\partial y^2} = 0 \tag{2.92}$$

Solutions of Laplace's equation having continuous first and second partial derivatives are called harmonic functions. Hence, the real and imaginary parts of an analytic function are harmonic functions, as seen with Eqs. 2.90 and 2.92.

Theorem 2.3 (Harmonic function) *If a function $f(z) = u + iv$ is analytic in a domain, its component functions u and v are harmonic in this domain.*

2.4.3 Laplace's Equation in Polar Form

For analytic functions expressed in polar coordinates, Laplace's equation in polar form (rather than Cartesian form) is of practical use. Differentiating both sides of the first Cauchy–Riemann equation in polar form (Eq. 2.83) with respect to r and both sides of the second with respect to θ yields

$$\begin{cases} r\dfrac{\partial^2 u}{\partial r^2} + \dfrac{\partial u}{\partial r} = \dfrac{\partial^2 v}{\partial r \partial \theta} \\[3mm] \dfrac{1}{r}\dfrac{\partial^2 u}{\partial \theta^2} = -\dfrac{\partial^2 v}{\partial r \partial \theta} \end{cases} \tag{2.93}$$

Adding these two equations gives

$$\frac{1}{r}\frac{\partial}{\partial r}\left(r\frac{\partial u}{\partial r}\right) + \frac{1}{r^2}\frac{\partial^2 u}{\partial \theta^2} = 0 \tag{2.94}$$

which is Laplace's equation in terms of u in polar form.

Similarly, differentiating both sides of the first Cauchy–Riemann equation with respect to θ and both sides of the second with respect to r yields

$$\begin{cases} \dfrac{\partial^2 u}{\partial r \partial \theta} = \dfrac{1}{r} \dfrac{\partial^2 v}{\partial \theta^2} \\[2ex] \dfrac{\partial^2 u}{\partial r \partial \theta} = -\dfrac{\partial v}{\partial r} - r \dfrac{\partial^2 v}{\partial r^2} \end{cases} \tag{2.95}$$

Subtracting the second equation from the first gives

$$\frac{1}{r} \frac{\partial}{\partial r} \left(r \frac{\partial v}{\partial r} \right) + \frac{1}{r^2} \frac{\partial^2 v}{\partial \theta^2} = 0 \tag{2.96}$$

which is Laplace's equation in terms of v in polar form.

2.4.4 Harmonic Conjugate

When two harmonic functions u and v satisfy the Cauchy–Riemann equations (Eq. 2.80 or 2.83) in a domain, v is said to be a harmonic conjugate of u in this domain. It is evident that if a function $f(z) = u + iv$ is analytic in a domain, v is a harmonic conjugate of u, and conversely, that if v is a harmonic conjugate of u in a domain, the function $f(z) = u + iv$ is analytic in this domain.

Theorem 2.4 (Harmonic conjugate) *A function $f(z) = u + iv$ is analytic in a domain if and only if v is a harmonic conjugate of u.*

Example 2.31 Let us consider real functions $u(x, y) = x^2 - y^2$ and $v(x, y) = 2xy$. Since these functions are respectively the real and imaginary parts of the analytic function $f(z) = z^2$, v is a harmonic conjugate of u.

In contrast, the function $f(z) = v + iu$ is not analytic, since the partial derivatives

$$\begin{cases} \partial v / \partial x = 2y \\ \partial v / \partial y = 2x \end{cases}$$

and

$$\begin{cases} \partial u / \partial y = -2y \\ -\partial u / \partial x = -2x \end{cases}$$

do not satisfy the Cauchy–Riemann equations, and thus, u is not a harmonic conjugate of v.

If v is a harmonic conjugate of u in some domain, it is not, in general, true that u is a harmonic conjugate of v in this domain. Instead, it is true that if v is a harmonic conjugate of u, $-u$ is a harmonic conjugate of v, that is, $f(z) = v - iu$ is analytic.

This can be understood by noting that $f(z) = u + iv$ is rewritten as $-if(z) = v - iu$ and that $f(z)$ is analytic if and only if $-if(z)$ is analytic.

Proposition 2.1 (Existence of a harmonic conjugate) *If a function u is harmonic, there exists a harmonic conjugate v such that $f = u + iv$ is an analytic function.*

Proof A harmonic conjugate v satisfies the Cauchy–Riemann equations $\partial v/\partial y = \partial u/\partial x$ and $\partial v/\partial x = -\partial u/\partial y$. Integrating $\partial v/\partial y = \partial u/\partial x$ with respect to y while keeping x constant gives

$$v = \int \frac{\partial u}{\partial x} dy + t(x)$$

where $t(x)$ is a real function of x. Substituting this into the second Cauchy–Riemann equation yields

$$\frac{\partial}{\partial x} \int \frac{\partial u}{\partial x} dy + t'(x) = -\frac{\partial u}{\partial y}$$

Since u is harmonic, differentiating this equation with respect to y results in Laplace's equation, revealing that a formula for $t'(x)$ involves x alone and $t(x)$ can be obtained by integration of $t'(x)$.   □

Example 2.32 Let us consider a real function $u(x, y) = y^3 - 3x^2y$. The second partial derivatives are

$$\begin{cases} \partial^2 u/\partial x^2 = -6y \\ \partial^2 u/\partial y^2 = 6y \end{cases}$$

which satisfy Laplace's equation $\partial u^2/\partial x^2 + \partial u^2/\partial y^2 = 0$ and the function u is harmonic. From the Cauchy–Riemann equations, it follows that

$$\begin{cases} \partial v/\partial y = \partial u/\partial x = -6xy \\ \partial v/\partial x = -\partial u/\partial y = -3y^2 + 3x^2 \end{cases}$$

Integrating the first equation with respect to y while keeping x constant gives

$$v = -3xy^2 + t(x)$$

where $t(x)$ is an arbitrary real function of x. Substituting this into the second equation above yields

$$-3y^2 + t'(x) = -3y^2 + 3x^2$$

which results in $t'(x) = 3x^2$, and thus, $t(x) = x^3 +$ constant. Hence, the harmonic conjugate of u is found as

$$v = -3xy^2 + x^3 + \text{constant}$$

and it follows that

$$f(z) = y^3 - 3x^2y + i(-3xy^2 + x^3 + \text{constant})$$
$$= iz^3 + \text{constant}$$

which is analytic.

2.5 Stream Function and Complex Potential

Finally, Solution Strategy is accomplished in this section. By virtue of Proposition 2.1, another mathematical function in addition to the velocity potential is constructed. It is shown that the function so created has the properties required to solve Motivating Problem 2.

2.5.1 Definition

The velocity potential Φ satisfies Laplace's equation, as shown in Chap. 1, and there-fore, is harmonic. From Proposition 2.1, it follows that there must exist a harmonic conjugate, denoted by Ψ, such that

$$\Omega = \Phi + i\Psi \qquad (2.97)$$

is analytic. The function Ψ is called the stream function, since Ψ is related to fluid streams as revealed in the later section. The analytic function Ω is called the complex potential.

Here, the properties of the velocity potential and stream function are sum-marized. As the vector function $\nabla\Phi$ is conservative, the vector function $\nabla\Psi$ is also conservative; thus, the properties associated with conservative fields, such as path independence of line integrals and the exactness of the differential form (Appendix C.2), hold for the stream function Ψ.

The complex potential Ω is analytic, and its component functions Φ and Ψ are harmonic, satisfying Laplace's equation:

$$\begin{cases} \nabla^2\Phi = 0 \\ \nabla^2\Psi = 0 \end{cases} \qquad (2.98)$$

Laplace's equation in terms of Φ is physically derived in Chap. 1. Although a physical derivation is also possible, Laplace's equation in terms of Ψ has immediately been derived through a mathematical manipulation.

In addition, Φ and Ψ are related through the Cauchy–Riemann equations

$$
\begin{cases}
\dfrac{\partial \Phi}{\partial x} = \dfrac{\partial \Psi}{\partial y} \\[2mm]
\dfrac{\partial \Phi}{\partial y} = -\dfrac{\partial \Psi}{\partial x}
\end{cases}
\tag{2.99}
$$

in Cartesian form or

$$
\begin{cases}
\dfrac{\partial \Phi}{\partial r} = \dfrac{1}{r}\dfrac{\partial \Psi}{\partial \theta} \\[2mm]
\dfrac{1}{r}\dfrac{\partial \Phi}{\partial \theta} = -\dfrac{\partial \Psi}{\partial r}
\end{cases}
\tag{2.100}
$$

in polar form.

As illustrated in Example 2.32, given a harmonic function (velocity potential Φ), the corresponding harmonic conjugate (stream function Ψ) can be obtained through the Cauchy–Riemann equations. In Chap. 1, the velocity potential for uniform flow and that for a source or sink are derived, to which the corresponding stream functions can be derived through the property of harmonic functions.

2.5.2 Uniform Flow

The velocity potential for uniform flow in the x direction is given by Eq. 1.55 as

$$
\Phi(x) = -\frac{q_u}{A}x + \Phi_0
\tag{2.101}
$$

From the Cauchy–Riemann equations in Cartesian form (Eq. 2.99), it follows that

$$
\begin{cases}
\dfrac{\partial \Psi}{\partial y} = \dfrac{\partial \Phi}{\partial x} = -\dfrac{q_u}{A} \\[2mm]
\dfrac{\partial \Psi}{\partial x} = -\dfrac{\partial \Phi}{\partial y} = 0
\end{cases}
\tag{2.102}
$$

Integrating the first equation with respect to y while keeping x constant gives

$$
\Psi = -\frac{q_u}{A}y + t(x)
\tag{2.103}
$$

where $t(x)$ is an arbitrary real function of x. Substituting this equation into the second equation of Eq. 2.102 gives $t'(x) = 0$ or $t(x) =$ constant. Hence, the stream function for uniform flow is obtained as

$$
\Psi = -\frac{q_u}{A}y + \Psi_0
\tag{2.104}
$$

where Ψ_0 is an arbitrary additive constant.

Substituting Eqs. 2.101 and 2.104 into Eq. 2.97 yields the complex potential for uniform flow

$$\Omega = -\frac{q_u}{A}z \tag{2.105}$$

where for simplicity a constant is omitted without essential loss of generality. This immediately shows that Ω is analytic in the entire complex plane.

2.5.3 Sources and Sinks

The velocity potential for sources and sinks is given by Eq. 1.58 as

$$\Phi(r) = -\frac{q_w}{2\pi h}\ln r + \Phi_0 \tag{2.106}$$

From the Cauchy–Riemann equations in polar form (Eq. 2.100), it follows that

$$\begin{cases} \dfrac{1}{r}\dfrac{\partial\Psi}{\partial\theta} = \dfrac{\partial\Phi}{\partial r} = -\dfrac{q_w}{2\pi h}\dfrac{1}{r} \\[2mm] \dfrac{\partial\Psi}{\partial r} = -\dfrac{1}{r}\dfrac{\partial\Phi}{\partial\theta} = 0 \end{cases} \tag{2.107}$$

Integrating the first equation with respect to θ while keeping r constant gives

$$\Psi = -\frac{q_w}{2\pi h}\theta + t(r) \tag{2.108}$$

where $t(r)$ is an arbitrary real function of r. Substituting this equation into the second equation of Eq. 2.107 gives $t'(r) = 0$ or $t(r) = $ constant. Hence, the stream function for sources and sinks is obtained as

$$\Psi = -\frac{q_w}{2\pi h}\theta + \Psi_0 \tag{2.109}$$

where Ψ_0 is an arbitrary additive constant.

Substituting Eqs. 2.106 and 2.109 into Eq. 2.97 yields

$$\Omega = -\frac{q_w}{2\pi h}(\ln r + i\theta) = -\frac{q_w}{2\pi h}\ln z \tag{2.110}$$

where for simplicity a constant is omitted without essential loss of generality. When a source or sink is located at $z_w = (x_w, y_w)$, the radius r is the distance between z and the source or sink, $r = |z - z_w|$ and the argument θ is the angle that $z - z_w$ (as a radius vector) makes with the positive real axis, $\theta = \arg(z - z_w)$. Hence, the complex potential for a source or sink at z_w can be written as

$$\Omega = -\frac{q_w}{2\pi h}\ln(z - z_w) \tag{2.111}$$

where a positive q_w corresponds to emanating flow from a source and a negative q_w corresponds to converging flow toward a sink. The complex potential Ω is analytic except at $z = z_w$ and on the branch cut.

Example 2.33 Let us confirm that the stream functions satisfy Laplace's equation. For uniform flow

$$\nabla^2 \psi = \frac{\partial^2 \psi}{\partial y^2} = -\frac{\partial}{\partial y} \frac{q_u}{A} = 0$$

For a source or sink, using Eq. A.12

$$\nabla^2 \psi = \frac{1}{r^2} \frac{\partial^2 \psi}{\partial \theta^2} = -\frac{1}{r^2} \frac{\partial}{\partial \theta} \frac{q_w}{2\pi h} = 0$$

The stream function for uniform flow and that for a source or sink indeed satisfy Laplace's equation.

2.5.4 Streamlines

Streamlines represent the paths of imaginary fluid particles in a flow domain and are defined as the instantaneous curves that are at every point tangent to the direction of the velocity at that point. Along a streamline, an element of arc $d\mathbf{r} = (dx, dy)$ is parallel to the velocity $\mathbf{V} = (V_x, V_y)$, as shown in Fig. 2.11.

The mathematical expression defining a streamline is therefore given by

$$\frac{dx}{V_x} = \frac{dy}{V_y} \tag{2.112}$$

or equivalently

$$V_y dx - V_x dy = 0 \tag{2.113}$$

where V_x and V_y are the velocity components in the x and y directions, such that a function $V = V_x + iV_y$ corresponds to the velocity vector $\mathbf{V}$.

Fig. 2.11 An element of arc $d\mathbf{r}$ along a streamline and the velocity vector $\mathbf{V}$

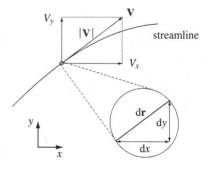

According to Darcy's law, Eq. 1.33 (or Eq. 1.24 in general), the velocity components are given as minus the gradient of Φ:

$$\begin{cases} V_x = -\dfrac{\partial \Phi}{\partial x} \\[2mm] V_y = -\dfrac{\partial \Phi}{\partial y} \end{cases} \tag{2.114}$$

Applying the Cauchy–Riemann equations (Eq. 2.99) to Eq. 2.114 yields another expression for the velocity components in terms of Ψ:

$$\begin{cases} V_x = -\dfrac{\partial \Psi}{\partial y} \\[2mm] V_y = \dfrac{\partial \Psi}{\partial x} \end{cases} \tag{2.115}$$

Now, substituting Eq. 2.115 into 2.113 gives

$$\frac{\partial \Psi}{\partial x}\mathrm{d}x + \frac{\partial \Psi}{\partial y}\mathrm{d}y = 0 \tag{2.116}$$

Since the exactness of the differential form (Eq. C.9 in Appendix C.2) holds for the stream function, the left-hand side of Eq. 2.116 is the exact differential form of Ψ; thus, along a streamline, it follows that

$$\mathrm{d}\Psi = 0 \tag{2.117}$$

Equation 2.117 implies that, in the same way as equipotential lines are obtained as curves of constant velocity potential Φ (Sect. 1.3.3), streamlines are obtained by setting Ψ equal to a constant in the equation

$$\Psi = \Psi(x, y) = \text{constant} \tag{2.118}$$

which describes a family of curves, for various values of the constant. That is, the level curves of Ψ are the streamlines.

Example 2.34 Figure 2.12 shows the streamlines of uniform flow given by Eq. 2.104 (or the imaginary part of Eq. 2.105) and a source or sink at z_w given by Eq. 2.109 (or the imaginary part of Eq. 2.111). For uniform flow, the streamlines are straight lines parallel to the flow direction which are all the same distance apart. For a source or sink, the streamlines are rays of constant θ emanating from or converging toward z_w.

Note that these figures are respectively drawn by contouring the values of Ψ computed within the domain. In Fig. 2.12b, a streamline is overlapped by the thick line from the source or sink in the negative x direction, which is caused by the jump on the contour plot and corresponds to the branch cut.

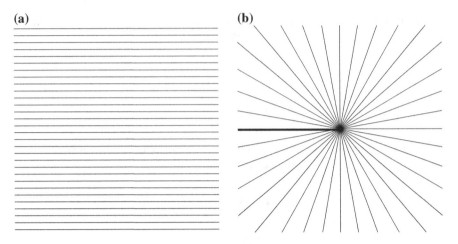

Fig. 2.12 Streamlines. **a** Uniform flow. **b** Source or sink

2.5.5 Complex Velocity

The velocity components V_x and V_y can be obtained by complex differentiation. From Eq. 2.78, the derivative of Ω with respect to z is given by

$$\frac{d\Omega}{dz} = \frac{\partial \Phi}{\partial x} + i\frac{\partial \Psi}{\partial x} = -V_x + iV_y \tag{2.119}$$

where Eqs. 2.114 and 2.115 are used. The complex velocity W is defined by

$$W = -\frac{d\Omega}{dz} = V_x - iV_y \tag{2.120}$$

from which the velocity components V_x and V_y are directly obtained. The magnitude of velocity $|\mathbf{V}|$ is also obtained as

$$|\mathbf{V}| = \sqrt{V_x^2 + V_y^2} = |W| \tag{2.121}$$

The velocity components in Cartesian and polar coordinates are related as

$$\begin{cases} V_x = V_r \cos\theta - V_\theta \sin\theta \\ V_y = V_r \sin\theta + V_\theta \cos\theta \end{cases} \tag{2.122}$$

where V_r and V_θ are the velocity components in the r and θ directions, respectively. Hence, the complex velocity in polar coordinates is given by

$$W = V_r \cos\theta - V_\theta \sin\theta - i(V_r \sin\theta + V_\theta \cos\theta)$$

$$= (V_r - iV_\theta)e^{-i\theta} \tag{2.123}$$

A point where both V_x and V_y vanish is called a stagnation point. From Eqs. 2.78 and 2.79, it follows that

$$W = -\frac{d\Omega}{dz} = -\frac{\partial \Phi}{\partial x} - i\frac{\partial \Psi}{\partial x} = i\frac{\partial \Phi}{\partial y} - \frac{\partial \Psi}{\partial y} = 0 \qquad (2.124)$$

at the stagnation point. Since $\partial \Psi/\partial x = 0$ and $\partial \Psi/\partial y = 0$, streamlines can intersect each other or abruptly change direction at the stagnation point.

Example 2.35 For uniform flow $\Omega = -(q_u/A)z$, the complex velocity is

$$W = -\frac{d\Omega}{dz} = \frac{q_u}{A}$$

Thus, $V_x = q_u/A$ and $V_y = 0$. There is no stagnation point in uniform flow. For a source at the origin $\Omega = -(q_w/2\pi h)\ln z$, the complex velocity is

$$W = -\frac{d\Omega}{dz} = \frac{q_w}{2\pi h}\frac{1}{z}$$

By using the exponential form $z = re^{i\theta}$, W becomes

$$W = \frac{q_w}{2\pi h}\frac{1}{r}e^{-i\theta} = (V_r - iV_\theta)e^{-i\theta}$$

Thus, $V_r = (q_w/2\pi h)/r$ and $V_\theta = 0$. The origin is a singular point, where W is indeterminate.

• **Solution to Task 2-1**

The complex potential for uniform flow with a source can be obtained by applying the principle of superposition to Eqs. 2.105 and 2.111. When the source with a flow rate q_s is located at z_s, the solution is

$$\Omega(z) = -\frac{q_u}{A}z - \frac{q_s}{2\pi h}\ln(z - z_s) \qquad (2.125)$$

Substituting $q_u/A = 1$, $q_s/h = 6$, and $z_s = -1$ into the solution yields

$$\Omega(z) = -z - \frac{3}{\pi}\ln(z + 1)$$

with which the corresponding complex potentials are computed. The real and imaginary parts of $\Omega(z)$ yield the velocity potential and stream function, respectively, and the equipotential lines and streamlines are obtained as shown in Fig. 2.13.

Since Eq. 1.65 is the real part of Eq. 2.125, the equipotential lines shown in Fig. 2.13a coincide with those in Fig. 1.5, which are of little use to trace the contaminants. On the other hand, the streamlines shown in Fig. 2.13b visually reveal

(a)   **(b)**

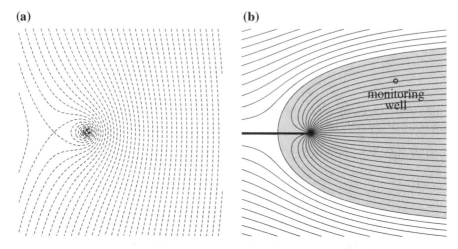

Fig. 2.13 Uniform flow with a source ($q_s/h = 6$). **a** Equipotential lines. **b** Streamlines. (As is the case with Example 2.34, the thick line emanating from the pollution source in the negative x direction corresponds to the branch cut and overlaps a streamline)

that the monitoring well is invaded by the contaminants from the pollution source. The stream function conveys useful information to draw exact flow paths.

The complex velocity is given by

$$W = -\frac{d\Omega}{dz} = 1 + \frac{3}{\pi}\frac{1}{z+1}$$

and from the solution of $W = 0$, a stagnation point is found at $z = -1 - 3/\pi = -1.955$, where the streamlines intersect each other and abruptly change direction, as seen in Fig. 2.13b.

The streamline passing through the stagnation point is called the dividing streamline (or water divide), which separates the fluid from uniform flow and the fluid from the pollution source. The monitoring well is inside the dividing streamline, and thus, the contaminants from the pollution source are detected at the monitoring point.

Let us further consider the dividing streamline. From Eq. 2.125, the stream function is given by

$$\Psi(z) = -\frac{q_u}{A}y - \frac{q_s}{2\pi h}\theta_s \tag{2.126}$$

where $\theta_s = \arg(z - z_s)$. For the streamlines in the upper half plane, the stagnation point is given by $y = 0$ and $\theta_s = \pi$; thus, the stream function at the stagnation point becomes $\Psi = -q_s/2h$. The equation for the dividing streamline is given by

$$-\frac{q_s}{2h} = -\frac{q_u}{A}y - \frac{q_s}{2\pi h}\theta_s \tag{2.127}$$

As x approaches ∞, θ_s approaches 0, and Eq. 2.127 gives

$$y = \frac{q_s/2h}{q_u/A} \tag{2.128}$$

which is the asymptote of the dividing streamline.

In a similar way, for the streamlines in the lower half plane, the stagnation point is given by $y = 0$ and $\theta_s = -\pi$; thus, the stream function becomes $\Psi = q_s/2h$. The equation for the dividing streamline is given by

$$\frac{q_s}{2h} = -\frac{q_u}{A}y - \frac{q_s}{2\pi h}\theta_s \tag{2.129}$$

and the asymptote of the dividing streamline is obtained as

$$y = -\frac{q_s/2h}{q_u/A} \tag{2.130}$$

The two asymptotes of the dividing streamline are given by

$$y = \pm\frac{q_s/2h}{q_u/A} \tag{2.131}$$

Substituting $q_u/A = 1$ and $q_s/h = 6$ into the equation yields $y = \pm 3$, the area between which at $x = \infty$ is contaminated.

• Solution to Task 2-2

The complex potential for uniform flow with a source and a sink can be obtained by applying the principle of superposition to Eqs. 2.105 and 2.111. When the source with a flow rate q_s is located at z_s and the sink with a flow rate q_e is at z_e, the solution is

$$\Omega(z) = -\frac{q_u}{A}z - \frac{q_s}{2\pi h}\ln(z - z_s) - \frac{q_e}{2\pi h}\ln(z - z_e) \tag{2.132}$$

Substituting $q_u/A = 1$, $q_s/h = 6$, $z_s = -1$, and $z_e = 1$ into the solution yields

$$\Omega(z) = -z - \frac{3}{\pi}\ln(z + 1) - \frac{q_e}{2\pi h}\ln(z - 1) \tag{2.133}$$

with which the corresponding complex potentials are computed. After a process of trial and error, it is found that $q_e/h = -3.5$ gives a possible solution to the extraction system, as shown in Fig. 2.14.

The complex velocity is given by

$$W = -\frac{d\Omega}{dz} = 1 + \frac{3}{\pi}\frac{1}{z + 1} - \frac{3.5}{2\pi}\frac{1}{z - 1}$$

(a)   **(b)**

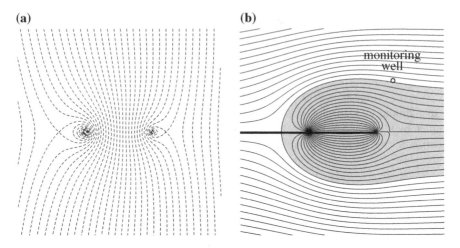

Fig. 2.14 Uniform flow with a source ($q_s/h = 6$) and a sink ($q_e/h = -3.5$). **a** Equipotential lines. **b** Streamlines. (The thick line emanating from the pollution source and that from the extraction well in the negative x direction correspond to the individual branch cuts and overlap streamlines)

and from the solution of $W = 0$, stagnation points are found at $z = -1.796$ and $z = 1.398$, as seen in Fig. 2.14b. The monitoring well is outside the dividing streamline and the streamlines emanating from the pollution source do not flow through the monitoring well. Hence, the contaminants are not detected at the monitoring point. Note that this is an approximate visual evaluation.

To evaluate analytically the minimum pumping rate to avoid the contamination at the monitoring well, the dividing streamline is of great use. As can be seen from Fig. 2.14, the dividing streamline originally emanates from the pollution source, then passes through and abruptly changes direction at the stagnation point. It is obvious that the area surrounded by the dividing streamline depends on the pumping rate of the extraction well; as the rate increases, the area shrinks. Hence, the dividing streamline that passes through the monitoring well corresponds to the required minimum pumping rate.

• Solution to Task 2-3
Taking the imaginary part of Eq. 2.133 yields the stream function

$$\Psi(z) = -y - \frac{3}{\pi}\theta_s - \frac{q_e}{2\pi h}\theta_e$$

where $\theta_s = \arg(z - z_s) = \arg(z + 1)$ and $\theta_e = \arg(z - z_e) = \arg(z - 1)$. For the streamlines in the upper half plane, the stagnation point is given by $y = 0$, $\theta_s = \pi$, and $\theta_e = \pi$; thus, the stream function at the stagnation point becomes

$$\Psi = 0 - \frac{3}{\pi}\pi - \frac{q_e}{2\pi h}\pi = -3 - \frac{1}{2}\frac{q_e}{h}$$

The minimum pumping rate can be obtained by equating this value and the stream function at the location of the monitoring well.

For the monitoring well, $y = 1.5$, $\theta_s = \arctan(1.5/2.5)$, and $\theta_e = \arctan(1.5/0.5)$; thus, the stream function becomes

$$-3 - \frac{1}{2}\frac{q_e}{h} = -1.5 - \frac{3}{\pi}\arctan(3/5) - \frac{q_e}{2\pi h}\arctan 3$$

which results in the analytical solution of $q_e/h = -3.267$. Figure 2.15 shows the equipotential lines and streamlines with $q_e/h = -3.267$. It is confirmed that the dividing streamline indeed passes through the monitoring well.

The complex velocity is given by

$$W = -\frac{d\Omega}{dz} = 1 + \frac{3}{\pi}\frac{1}{z+1} - \frac{3.267}{2\pi}\frac{1}{z-1}$$

and from the solution of $W = 0$, stagnation points are found at $z = -1.806$ and $z = 1.371$, as seen in Fig. 2.15b. The monitoring well is exactly on the streamline that emanates from the pollution source, and thus, $q_e/h = -3.267$ is the minimum value for the required flow rate.

From Eq. 2.132, the stream function for uniform flow with a source and a sink is given by

$$\Psi(z) = -\frac{q_u}{A}y - \frac{q_s}{2\pi h}\theta_s - \frac{q_e}{2\pi h}\theta_e \tag{2.134}$$

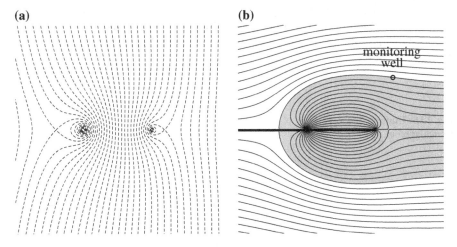

(a)   **(b)**

Fig. 2.15 Uniform flow with a source ($q_s/h = 6$) and a sink ($q_e/h = -3.267$). **a** Equipotential lines. **b** Streamlines

With the same argument as before, it can be shown that the equations for the dividing streamlines are given by

$$\mp \left(\frac{q_s}{2h} + \frac{q_e}{2h}\right) = -\frac{q_u}{A}y - \frac{q_s}{2\pi h}\theta_s - \frac{q_e}{2\pi h}\theta_e \qquad (2.135)$$

As x approaches ∞, θ_s and θ_e approach 0, and the two asymptotes of the dividing streamline are obtained as

$$y = \pm\frac{q_s/2h + q_e/2h}{q_u/A} \qquad (2.136)$$

Substituting $q_u/A = 1$, $q_s/h = 6$, and $q_e/h = -3.267$ into the equation yields $y = \pm 1.367$, the area between which at $x = \infty$ is contaminated.

Motivating Problem 3: Groundwater Flow Over a Circular Pillar

Groundwater flows in the x direction with a uniform flow velocity $q_u/A = 1$. For some construction purpose, an impermeable circular pillar of radius $R = 1$ is installed through the flow medium, as shown in Fig. 2.16.

Task 3-1 Draw the flow profile and discuss the effect of the circular pillar on the groundwater flow.

Task 3-2 Evaluate the discharge profile along the y axis.

• Solution Strategy to Motivating Problem 3

At first glance, Motivating Problem 3 may appear totally different from Motivating Problem 2, where uniform flow with a source and a sink is considered and no obstacle to flow is dealt with. This is not necessarily so.

Further insights into the properties of complex potential broaden the interpretation of streamlines and the usage of stream function. Indeed, a slight modification of the solution to Motivating Problem 2 gives the solution to the current problem.

Fig. 2.16 Uniform flow over a circular pillar

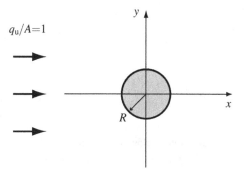

2.6 Further Topics in Complex Potential

To broaden its practical applicability, additional aspects of complex potential are discussed in this section, including the orthogonal families of curves, streamlines as impermeable boundaries, and the discharge evaluation with the stream function.

2.6.1 Orthogonal Families

Let us consider two families of curves

$$\begin{cases} u(x, y) = p \\ v(x, y) = s \end{cases} \tag{2.137}$$

where u and v are the real and imaginary parts of an analytic function $f(z)$ and p and s are any constants.

Let $u(x, y) = p_1$ and $v(x, y) = s_1$, where p_1 and s_1 are particular constants, be any two members of the respective families. Differentiating $u(x, y) = p_1$ with respect to x gives

$$\frac{\partial u}{\partial x} + \frac{\partial u}{\partial y} \frac{dy}{dx} = 0 \tag{2.138}$$

Then the slope of $u(x, y) = p_1$ is

$$\frac{dy}{dx} = -\frac{\partial u}{\partial x} \bigg/ \frac{\partial u}{\partial y} \tag{2.139}$$

In a similar way, the slope of $v(x, y) = s_1$ is obtained as

$$\frac{dy}{dx} = -\frac{\partial v}{\partial x} \bigg/ \frac{\partial v}{\partial y} \tag{2.140}$$

The product of the slopes becomes

$$\left(\frac{\partial u}{\partial x}\frac{\partial v}{\partial x}\right) \bigg/ \left(\frac{\partial u}{\partial y}\frac{\partial v}{\partial y}\right) = -\left(\frac{\partial v}{\partial y}\frac{\partial u}{\partial y}\right) \bigg/ \left(\frac{\partial u}{\partial y}\frac{\partial v}{\partial y}\right) = -1 \tag{2.141}$$

where the Cauchy–Riemann equations are used. This indicates that, if a function $f(z) = u + iv$ is analytic, the families of curves given by Eq. 2.137 are orthogonal. This is true only if the partial derivatives in Eq. 2.141 are not equal to zero.

Since the complex potential $\Omega = \Phi + i\Psi$ is analytic, it follows that two families of curves, equipotential lines and streamlines, defined by

$$\begin{cases} \Phi(x, y) = p \\ \Psi(x, y) = s \end{cases} \tag{2.142}$$

are orthogonal. Each curve of equipotential lines is perpendicular to each curve of streamlines at the point of intersection. This is true only if the partial derivatives of Φ and Ψ with respect to x and y are not equal to zero, that is, the complex velocity W is not equal to zero.

As discussed in Sect. 2.5.5, when $W = 0$, the point is a stagnation point. If the point of intersection is a stagnation point, the equipotential line and the streamline may not be mutually orthogonal. Instead, streamlines may intersect each other or abruptly change direction.

Example 2.36 The complex potential for uniform flow given by Eq. 2.105 has two families of curves, equipotential lines and streamlines

$$
\begin{cases}
\Phi(x, y) = -\dfrac{q_u}{A}x = p \\[2mm]
\Psi(x, y) = -\dfrac{q_u}{A}y = s
\end{cases}
$$

which are respectively parallel to y and x axes, and thus obviously orthogonal, as shown in Fig. 2.17a.

Similarly, the complex potential for a source or sink given by Eq. 2.111 has the following two families:

$$
\begin{cases}
\Phi(x, y) = -\dfrac{q_w}{2\pi h}\ln r = -\dfrac{q_w}{4\pi h}\ln\left(x^2 + y^2\right) = p \\[2mm]
\Psi(x, y) = -\dfrac{q_w}{2\pi h}\theta = -\dfrac{q_w}{2\pi h}\arctan\dfrac{y}{x} = s
\end{cases}
$$

(a)   **(b)**

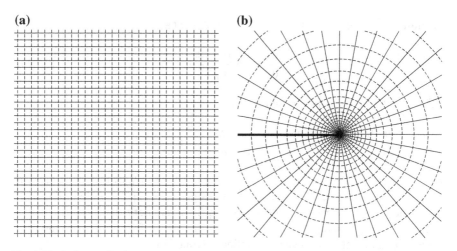

Fig. 2.17 Orthogonality between equipotential lines and streamlines. **a** Uniform flow. **b** Source or sink

(a)   **(b)**

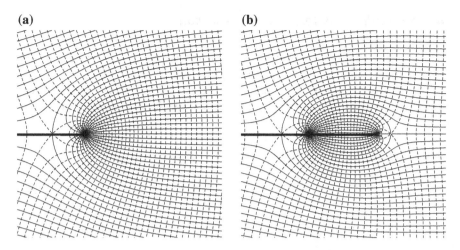

Fig. 2.18 Orthogonality between equipotential lines and streamlines of Motivating Problem 2. **a** Contamination. **b** Extraction

and the slopes dy/dx of $\Phi = p_1$ and $\Psi = s_1$ are respectively obtained as $-x/y$ and y/x, which implies the two families are orthogonal (Fig. 2.17b).

Example 2.37 The orthogonality also holds for superposition of different types of flow as long as the resultant flow is defined by analytic functions. Figure 2.18 shows the orthogonality between the equipotential lines and streamlines for uniform flow with a source and a sink (Motivating Problem 2). Note that at stagnation points, $z = -1.955$ in Fig. 2.18a and $z = -1.806$ and 1.371 in Fig. 2.18b, the equipotential lines and streamlines are not mutually orthogonal and the streamlines abruptly change direction.

As discussed in Sect. 1.2, many physical problems are governed by Laplace's equation, and thus, the solutions are given by a harmonic function Φ and its harmonic conjugate Ψ. For individual physical processes, the orthogonality between Φ and Ψ holds. Table 2.1 summarizes the orthogonal families for different physical phenomena.

Table 2.1 Orthogonal families for physical processes

Physical process	$\Phi(x, y)$	$\Psi(x, y)$
Fluid flow	Equipotential lines	Streamlines
Fickian diffusion	Concentration	Lines of solute flow
Heat conduction	Isotherms	Heat flow lines
Gravitational fields	Gravitational potential	Lines of force
Electrostatic fields	Equipotential lines	Lines of electrical force

2.6.2 Streamlines as Impermeable Boundaries

It follows from the orthogonality that the velocity potential is constant in the direction normal to the streamlines, across which no flow occurs. Hence, the streamlines can be interpreted as impermeable boundaries. In Figs. 2.13, 2.14, and 2.15, for instance, the shaded areas may be interpreted as $\subset$-shaped obstacles instead of contaminated areas.

For uniform flow with a source and a sink given by Eq. 2.132, if the flow rates are balanced, $q_s = -q_e$ and $z_s = -z_e = -d$, then

$$\Omega(z) = -\frac{q_u}{A}z - \frac{q_s}{2\pi h}\ln\frac{z+d}{z-d} \tag{2.143}$$

Figure 2.19 shows the streamlines given by Eq. 2.143. It is seen that the dividing streamline forms an oval contour.

Although Eq. 2.143 is derived for uniform flow around a source-sink pair, the resultant flow can be interpreted as the flow over an oval object, which is known as Rankine's oval. Inside the oval contour, the fluid emanates from the source and flows into the sink, whereas the fluid outside the oval is attributed to uniform flow.

The stagnation points are obtained from the solution of

$$W = -\frac{d\Omega}{dz} = \frac{q_u}{A} + \frac{q_s}{2\pi h}\left(\frac{1}{z+d} - \frac{1}{z-d}\right) = 0 \tag{2.144}$$

which yields

$$z = \pm\sqrt{d^2 + \frac{A}{q_u}\frac{q_s}{\pi h}d} \tag{2.145}$$

Fig. 2.19 Uniform flow over Rankine's oval

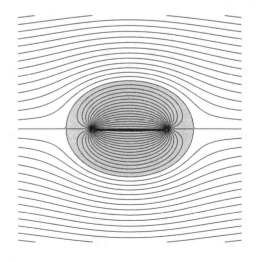

There exist two stagnation points, one to the left of the source and the other to the right of the sink, both on the x axis (Fig. 2.19).

It is deduced that the shape of Rankine's oval depends on the distance between the source and the sink, $2d$, and the ratio between the flow rates, q_s/q_u. As the source and the sink get closer to each other (d approaches zero), while $(q_s/q_u)d$ is held constant, the oval shape becomes more and more circular.

The coalescence of the source and the sink, with the flow rates equal in magnitude but opposite in sign, at a single point forms a dipole. By letting d approach zero and $m = q_s/(2\pi h)$ approach ∞ in such a way that $2dm = \sigma$ is finite, the second term on the right-hand side of Eq. 2.143 yields

$$
\begin{aligned}
-\lim_{d \to 0} \frac{q_s}{2\pi h} \ln \frac{z+d}{z-d} &= -\lim_{d \to 0} m \left[\ln(z+d) - \ln(z-d)\right] \\
&= -\lim_{d \to 0} 2dm \frac{\ln(z+d) - \ln(z-d)}{2d} \\
&= -\sigma \frac{d}{dz} \ln z = -\frac{\sigma}{z}
\end{aligned}
\tag{2.146}
$$

which represents flow caused by a dipole. Then, the complex potential Eq. 2.143 becomes

$$
\Omega(z) = -\frac{q_u}{A} z - \frac{\sigma}{z}
\tag{2.147}
$$

which expresses flow over a circular object.

By using the exponential form $z = re^{i\theta}$, the stream function is found to be

$$
\Psi = \mathrm{Im}\, \Omega(r, \theta) = -\mathrm{Im}\left[\frac{q_u}{A} re^{i\theta} + \frac{\sigma}{r} e^{-i\theta}\right] = -\left(\frac{q_u}{A} r - \frac{\sigma}{r}\right) \sin\theta
\tag{2.148}
$$

which becomes zero when $r = \sqrt{\sigma/(q_u/A)}$ or $\sin\theta = 0$. Hence, the streamline $\Psi = 0$ consists of the circle

$$
|z| = \sqrt{\frac{\sigma}{q_u/A}}
\tag{2.149}
$$

and the x axis ($\theta = 0$ and $\theta = \pi$). This implies that the streamline on the x axis (beyond the stagnation points) passes along the circumference of the circle and confirms that Rankine's oval becomes a circle in the limit of $d \to 0$.

It should be noted that the stagnation points given by Eq. 2.145 approach

$$
\begin{aligned}
z &= \pm \lim_{d \to 0} \sqrt{d^2 + \frac{A}{q_u} \frac{q_s}{\pi h} d} = \pm \lim_{d \to 0} \sqrt{d^2 + \frac{2dm}{q_u/A}} \\
&= \pm \sqrt{\frac{\sigma}{q_u/A}}
\end{aligned}
\tag{2.150}
$$

which are on the circle. The same result follows from the solution of $W = 0$. From Eq. 2.147, the corresponding complex velocity is

$$W = \frac{q_u}{A} - \frac{\sigma}{z^2} \tag{2.151}$$

and stagnation points are found on the circular object at

$$z = \pm\sqrt{\frac{\sigma}{q_u/A}} \tag{2.152}$$

The distance between the two is the diameter of the circle, and thus the radius is $\sqrt{\sigma/(q_u/A)}$. Let R be the radius of the circle, then $\sigma = (q_u/A)R^2$ and Eq. 2.147 becomes

$$\Omega(z) = -\frac{q_u}{A}\left(z + \frac{R^2}{z}\right) \tag{2.153}$$

for flow over a circular object of radius R.

• **Solution to Task 3-1**

The complex potential for uniform flow over a circular pillar is given by Eq. 2.153. Substituting $q_u/A = 1$ and $R = 1$ into the solution yields

$$\Omega(z) = -z - \frac{1}{z} \tag{2.154}$$

with which the corresponding complex potentials are computed. Figure 2.20 shows the equipotential lines and streamlines.

The complex velocity is given by

$$W = -\frac{d\Omega}{dz} = 1 - \frac{1}{z^2} \tag{2.155}$$

and from the solution of $W = 0$, stagnation points are found at $z = \pm 1$, as expected. Flow outside the circumference of the pillar is caused by uniform flow only. Away from the object, flow is essentially uniform as can be seen by evenly spaced horizontal streamlines. When approaching the pillar, flow is gradually distorted and the streamlines detour and become denser near the circumference of the pillar.

2.6.3 Discharge and Stream Function

Let us consider a discharge $\Delta q/h$ (in terms of volume per unit thickness normal to the xy plane per unit time) flowing in the channel between two streamlines with stream functions of Ψ_1 and Ψ_2, respectively, as shown in Fig. 2.21. This discharge is constant in the channel because flow cannot leave across the streamlines.

(a) **(b)**

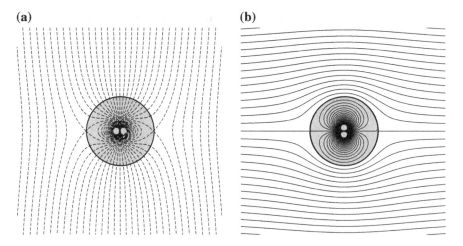

Fig. 2.20 Uniform flow over an impermeable circular object of radius $R = 1$. **a** Equipotential lines. **b** Streamlines. (In the vicinity of the center of the object, equipotential lines and streamlines become too dense and some of them are not plotted)

Fig. 2.21 Discharge $\Delta q/h$ between two streamlines with stream functions of Ψ_1 and Ψ_2

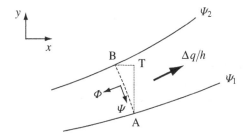

A temporary point T is set so that the segment AT is parallel to the y axis and the segment TB is parallel to the x axis. By continuity of flow, the discharge must pass through ATB, and thus

$$\frac{\Delta q}{h} = \int_A^T V_x \mathrm{d}y + \int_B^T V_y \mathrm{d}x \tag{2.156}$$

where V_x and V_y are given by Eq. 2.115; thus

$$\begin{aligned}
\frac{\Delta q}{h} &= -\int_A^T \frac{\partial \Psi}{\partial y}\mathrm{d}y + \int_B^T \frac{\partial \Psi}{\partial x}\mathrm{d}x \\
&= -\Psi_T + \Psi_A + \Psi_T - \Psi_B = \Psi_A - \Psi_B \\
&= \Psi_1 - \Psi_2
\end{aligned} \tag{2.157}$$

where Ψ_A, Ψ_B, and Ψ_T are the stream functions at A, B, and T, respectively.

It follows that the discharge $\Delta q/h$ between two streamlines is given as the difference in the values of stream function corresponding to those streamlines. It should be noted that the path used for the integration is immaterial, since path independence of line integrals holds for the stream function Ψ. That is, the result depends only on Ψ_1 and Ψ_2.

Since Φ increases in the direction against the flow direction and Φ and Ψ form a Cartesian coordinate system, Eq. 2.157 can be written as

$$\frac{\Delta q}{h} = -\Delta\Psi \tag{2.158}$$

The dimension of stream function is that of discharge, volume per unit thickness per unit time.

The same result can be derived through the definition of the discharge, Eq. 1.15, across a curve C. Applying Eq. 2.115 to Eq. 1.15, it follows that

$$Q = \int_C (V_x dy - V_y dx) = \int_C \left(-\frac{\partial\Psi}{\partial y} dy - \frac{\partial\Psi}{\partial x} dx \right)$$

$$= -\int_C d\Psi \tag{2.159}$$

which states that the discharge Q can be obtained by using the integral of the stream function along the curve C. For the points A and B in Fig. 2.21, consider the connecting curve C; then it follows that

$$Q = -\int_C d\Psi = -\int_A^B d\Psi = \Psi_A - \Psi_B = \Psi_1 - \Psi_2 \tag{2.160}$$

which is consistent with Eq. 2.157.

When C is a closed curve (A = B) and there is no singularity interior to C, the discharge becomes zero, since path independence of line integrals holds for Ψ and Eq. 2.159 yields $Q = \Psi_A - \Psi_B = 0$. Conversely, if there exist singularities interior to C, the discharge across C may not be zero.

Example 2.38 Let us consider a source with a flow rate per unit thickness q_w/h at the origin, the complex potential of which is given by

$$\Omega = -\frac{q_w}{2\pi h}\ln z = -\frac{q_w}{2\pi h}\ln r - i\frac{q_w}{2\pi h}\theta \tag{2.161}$$

For the stream function $\Psi = -(q_w/2\pi h)\theta$, it follows that

$$\frac{\partial\Psi}{\partial\theta} = -\frac{q_w}{2\pi h}$$

From Eq. 2.159, the discharge across a circle C of radius r_0 with its center at the origin, $|z| = r_0$, is given by

$$Q = - \oint_C d\Psi = \int_0^{2\pi} \frac{q_w}{2\pi h} d\theta = \frac{q_w}{h}$$

which is independent of r_0 and equal to the flow rate per unit thickness of the source.

It should be noted that the nonzero discharge across the circle enclosing the source does not contradict the condition of incompressibility, Eq. 1.17. The source is the singularity at which a fluid of discharge q_w/h is introduced. If a closed curve does not enclose the source, the discharge becomes zero.

2.6.4 Circulation and Velocity Potential

Let us consider a circulation Γ along a curve C connecting points A and B. Applying Eq. 2.114 to the definition of the circulation, Eq. 1.19, it follows that

$$\Gamma = \int_C \left(V_x dx + V_y dy \right) = \int_C \left(-\frac{\partial \Phi}{\partial x} dx - \frac{\partial \Phi}{\partial y} dy \right)$$

$$= - \int_C d\Phi \tag{2.162}$$

which states that the circulation Γ can be obtained by using the integral of the velocity potential along the curve C.

When C is a closed curve (A = B) and there is no singularity interior to C, the circulation becomes zero, since path independence of line integrals holds for Φ and Eq. 2.162 yields $\Gamma = \Phi_A - \Phi_B = 0$. Conversely, if there exist singularities interior to C, the circulation along C may not be zero.

Example 2.39 Let us consider a source with a flow rate per unit thickness q_w/h at the origin, expressed by Eq. 2.161. For the velocity potential $\Phi = -(q_w/2\pi h) \ln r$, it follows $\partial \Phi / \partial \theta = 0$. From Eq. 2.162, the circulation along a circle C is given by

$$\Gamma = - \oint_C d\Phi = - \int_0^{2\pi} 0 d\theta = 0$$

which satisfies the condition of irrotationality.

• Solution to Task 3-2
The stream function for uniform flow over a circular pillar of radius R is the imaginary part of Eq. 2.153, given by

$$\Psi(x, y) = -\frac{q_u}{A} \left(y - \frac{R^2 y}{x^2 + y^2} \right)$$

Fig. 2.22 Discharge (*bars*) and velocity (*dashed lines*) of uniform flow over an impermeable circular object of radius $R = 1$

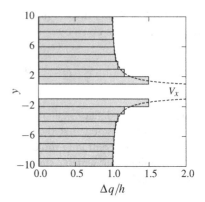

Substituting $q_u/A = 1, R = 1$, and $x = 0$ into the solution yields the stream function along the y axis:

$$\Psi(0, y) = -y + \frac{1}{y} \tag{2.163}$$

From Eq. 2.158, the discharge $\Delta q/h$ flowing through an interval Δy can be evaluated by using the difference in the values of stream function. Figure 2.22 shows the values of $\Delta q/h$ for $\Delta y = 1$ along the y axis.

In the limit of $\Delta y \to 0$, the discharge $\Delta q/h$ passing through Δy reduces to

$$\lim_{\Delta y \to 0} \frac{\Delta q/h}{\Delta y} = -\lim_{\Delta y \to 0} \frac{\Delta \Psi}{\Delta y} = -\frac{\partial \Psi}{\partial y}$$

which is equal to V_x, and from Eq. 2.163, it follows that

$$V_x = 1 + \frac{1}{y^2} \tag{2.164}$$

along the y axis.

The same result can be directly obtained from the complex velocity, Eq. 2.155, which gives

$$\begin{cases} V_x = 1 - \dfrac{x^2 - y^2}{(x^2 + y^2)^2} \\ V_y = -\dfrac{2xy}{(x^2 + y^2)^2} \end{cases}$$

and it follows that

$$\begin{cases} V_x = 1 + \dfrac{1}{y^2} \\ V_y = 0 \end{cases}$$

along the y axis.

The profile of V_x given by Eq. 2.164 is shown in Fig. 2.22. The velocity takes its maximum $V_x = 2$ on the circumference of the object at $y = \pm 1$ and decreases as $|y|$ increases. In the limit of $|y| \to \infty$, V_x approaches 1, which is equal to the uniform flow velocity $q_u/A = 1$.

2.6.5 Dipoles

The dipole derived in Eq. 2.146 is oriented in the x direction. The orientation of the dipole depends on the direction from which the source approaches the sink. Let us consider the source approaching the sink in the direction making an angle δ with the x axis. Substituting $de^{i\delta}$ for d into Eq. 2.146 yields the complex potential

$$\Omega = -\frac{\sigma e^{i\delta}}{z} \tag{2.165}$$

for a dipole oriented at an angle δ to the x direction.

Figure 2.23 shows the equipotential lines and streamlines for a dipole with $\delta = \pi/6$. The curves are families of circles through the dipole with their centers on mutually orthogonal lines.

Example 2.40 In the solution to Task 3-1, let us consider a dipole oriented against uniform flow, that is, $\delta = \pi$ in Eq. 2.165. Then the complex potential for uniform flow with a dipole oriented against it becomes

$$\Omega = -z + \frac{1}{z} \tag{2.166}$$

(a)   **(b)**

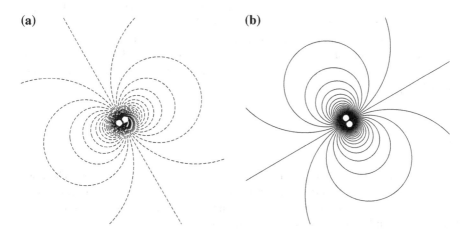

Fig. 2.23 A dipole in the direction making an angle $\pi/6$ with the x direction. **a** Equipotential lines. **b** Streamlines. (In the vicinity of the origin, equipotential lines and streamlines become too dense and some of them are not plotted)

(a)   **(b)**

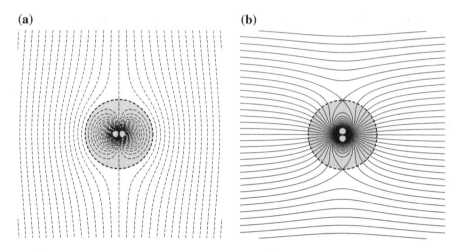

Fig. 2.24 Uniform flow with a dipole oriented against the flow. **a** Equipotential lines. **b** Streamlines

with which the corresponding equipotential lines and streamlines are obtained as shown in Fig. 2.24.

The circumference of the circular object coincides with an equipotential line, which indicates that the circular pillar becomes an equipotential object rather than an obstacle to flow. Consequently, the streamlines are perpendicular to the circular object. Inflow occurs along the upstream (left-hand) side of the circle and outflow along the downstream (right-hand) side.

The complex velocity is given by

$$W = -\frac{d\Omega}{dz} = 1 + \frac{1}{z^2}$$

and from the solution of $W = 0$, stagnation points are found at $z = \pm i$, where the streamlines intersect each other, as seen in Fig. 2.24b. Away from the equipotential object, flow is essentially uniform as can be seen by evenly spaced horizontal streamlines. When approaching the circular object, flow is gradually distorted and the streamlines become perpendicular to the circular equipotential line, except at the stagnation points.

2.6.6 Vortices

A vortex is a point around which a fluid flows along concentric circles. The corresponding complex potential is given by interchanging the velocity potential and stream function for a source or sink; mathematically, the complex potential for a vortex at z_v can be obtained by dividing that for a source or sink by an imaginary unit i

$$\Omega = -\frac{\gamma}{2\pi i} \ln(z - z_v) \tag{2.167}$$

(a)   **(b)**

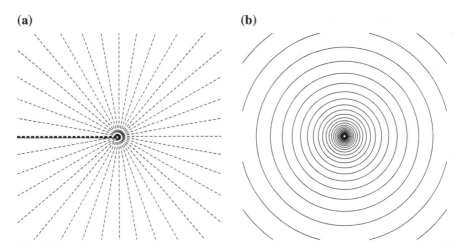

Fig. 2.25 Flow caused by a vortex. **a** Equipotential lines. (The thick line emanating from the vortex in the negative x direction corresponds to the branch cut and overlaps an equipotential line). **b** Streamlines

where γ is the strength of the vortex. The resultant flow caused by the vortex is shown in Fig. 2.25. The equipotential lines are rays emanating from the vortex and the streamlines are concentric circles centered at the vortex.

By using the exponential form $z - z_v = re^{i\theta}$, it follows that

$$\Omega = -\frac{\gamma}{2\pi i}(\ln r + i\theta) = -\frac{\gamma}{2\pi}\theta + i\frac{\gamma}{2\pi}\ln r \qquad (2.168)$$

For the stream function $\Psi = (\gamma/2\pi)\ln r$, it follows that $\partial\Psi/\partial\theta = 0$. From Eq. 2.159, the discharge across a circle C with its center at z_v is given by

$$Q = -\oint_C d\Psi = \int_0^{2\pi} 0 \, d\theta = 0 \qquad (2.169)$$

which satisfies the condition of incompressibility.

For the velocity potential $\Phi = -(\gamma/2\pi)\theta$, it follows that

$$\frac{\partial\Phi}{\partial\theta} = -\frac{\gamma}{2\pi} \qquad (2.170)$$

From Eq. 2.162, the circulation along a circle C enclosing $z = z_v$ is given by

$$\Gamma = -\oint_C d\Phi = \int_0^{2\pi} \frac{\gamma}{2\pi} d\theta = \gamma \qquad (2.171)$$

which is equal to the strength of the vortex.

The nonzero circulation along the circle enclosing the vortex does not contradict the condition of irrotationality, Eq. 1.21. The vortex is the singularity at which a circulation of γ is introduced. If a closed curve does not enclose the vortex, the circulation becomes zero.

The complex velocity is given by

$$W = -\frac{d\Omega}{dz} = \frac{\gamma}{2\pi i}\frac{1}{z - z_v} \tag{2.172}$$

and by using the exponential form, W becomes

$$W = \frac{\gamma}{2\pi i}\frac{1}{r}e^{-i\theta} = -i\frac{\gamma}{2\pi}\frac{1}{r}e^{-i\theta} = (V_r - iV_\theta)e^{-i\theta} \tag{2.173}$$

Thus, $V_r = 0$ and $V_\theta = (\gamma/2\pi)/r$. The direction of the velocity is tangential to the concentric circles and the magnitude is inversely proportional to the distance from the vortex. For positive values of γ, flow becomes counterclockwise, while for negative values of γ, flow is clockwise.

Example 2.41 In the solution to Task 3-1, let us consider uniform flow over a circular object with circulation. By adding a vortex at the origin to Eq. 2.154, it follows that

$$\Omega = -z - \frac{1}{z} - \frac{\gamma}{2\pi i}\ln z$$

Since the vortex has concentric circular streamlines, the addition of the vortex to flow over a circular object does not deform the object. For $\gamma = -2\pi$, the corresponding equipotential lines and streamlines are obtained, as shown in Fig. 2.26.

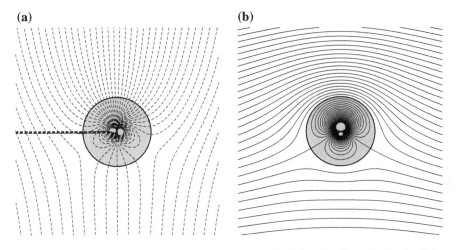

(a)   **(b)**

Fig. 2.26 Uniform flow over an impermeable circular object of radius $R = 1$ with circulation $\gamma = -2\pi$. **a** Equipotential lines. **b** Streamlines

(a)   **(b)**

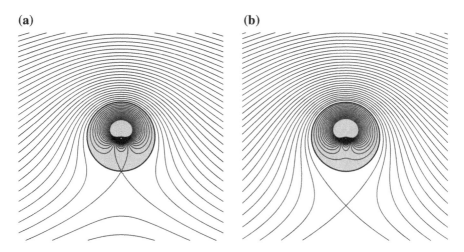

Fig. 2.27 Uniform flow over an impermeable circular object of radius $R = 1$ with circulation. **a** Circulation of $\gamma = -4\pi$. **b** Circulation of $\gamma = -5\pi$

Flow caused by the vortex is in the clockwise direction, and thus, flow above the circle is accelerated, while flow below the circle is decelerated, as can be confirmed from the dense streamlines above and the sparse streamlines below the circle.

The complex velocity is given by

$$W = -\frac{d\Omega}{dz} = 1 - \frac{1}{z^2} + \frac{\gamma}{2\pi i}\frac{1}{z}$$

and from the solution of $W = 0$, stagnation points are found at

$$z = \frac{(\gamma/2\pi)i \pm \sqrt{4 - (\gamma/2\pi)^2}}{2}$$

If $0 \le |\gamma/2\pi| < 2$, there exist two stagnation points on the unit circle $|z| = 1$. For $\gamma = -2\pi$, for instance, the stagnation points are found at $z = \pm\sqrt{3}/2 - i/2$ on the circle, as seen in Fig. 2.26. If $|\gamma/2\pi| = 2$, there is one stagnation point either at $z = +i$ or $z = -i$ on the unit circle. If $|\gamma/2\pi| > 2$, stagnation points lie on the imaginary axis, one outside the unit circle and the other inside the circle, which has no physical meaning. Figure 2.27 shows such flow profiles for $\gamma = -4\pi$ and $\gamma = -5\pi$.

The profiles depend on the strength of the vortex γ, and the stagnation points are found in accordance with the discussion above. For $\gamma = -4\pi$, the stagnation point is at $z = -i$ on the object and for $\gamma = -5\pi$, the stagnation points are at $z = -2i$ in the flow domain and at $z = -0.5i$ inside the object.

Chapter 3
Transformation and Conformal Mapping

Motivating Problem 4: Groundwater Flow Over a Thin Object

Groundwater flows in the direction making an angle $\delta = \pi/6$ with the x direction with a uniform flow velocity $q_u/A = 1$. For some construction purposes, an impermeable thin shield of length $4R = 4$ is installed through the flow medium, as shown in Fig. 3.1.

Task 4-1 Draw the flow profile and discuss the effect of the thin shield on the groundwater flow.

Task 4-2 Evaluate the velocity profiles along the x and y axes and discuss the velocity variation in the flow domain.

Task 4-3 Redo the tasks for the case when the thin object is an equipotential line segment, such as an infinite conductivity fracture.

Fig. 3.1 Uniform inclined flow over a thin object

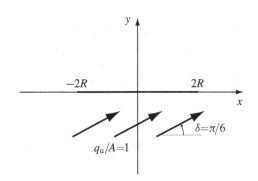

• Solution Strategy to Motivating Problem 4

By comparing Motivating Problems 3 and 4, differences are found in the direction of uniform flow and the shape of the obstacle. To find the solution to Motivating Problem 4, a geometric modification of the solution to Motivating Problem 3 is of help.

The geometric approach in complex analysis used here is known as conformal mapping, which, when applicable, is invaluable to engineering practitioners as an aid in solving physical problems.

3.1 Analytic Function and Mapping

In this section, a geometric interpretation of a complex function as a mapping is introduced. Exploring geometric properties of general transformations provides new insights into complex functions. In particular, the concept of conformal mapping is profoundly discussed in relation to analytic functions.

3.1.1 Analyticity and Conformality

A function f of a complex variable z, given by

$$w = f(z) = f(x + iy) = u(x, y) + iv(x, y) \tag{3.1}$$

defines a mapping or transformation that establishes a correspondence between points in the w and z planes. For any point z_0, the point $w_0 = f(z_0)$ is called the image of z_0 with respect to f.

Let us consider a curve C in the z plane, represented by the parametric equation

$$z = z(t) = x(t) + iy(t) \tag{3.2}$$

where t is the parameter. The equation $w(t) = f(z(t))$ is a parametric representation of the image C' of C under the mapping f.

A vector tangent to the curve C at the point $z_0 = z(t_0)$ is given by $z'(t_0)$, where the complex number $z'(t_0)$ is expressed as a vector. The angle θ of the vector with respect to the x axis is

$$\theta = \arg z'(t_0) \tag{3.3}$$

Similarly, a vector tangent to the image curve C' in the w plane at the point $w_0 = w(t_0) = f(z(t_0))$ is given by $w'(t_0) = f'(z(t_0))$. Using the chain rule, it follows that

$$w'(t_0) = f'(z_0)z'(t_0) \tag{3.4}$$

provided that $f(z)$ is analytic at $z = z_0$.

The angle of the vector with respect to the u axis is

$$\arg w'(t_0) = \arg f'(z_0) + \arg z'(t_0) \tag{3.5}$$

where the identity Eq. 2.34 is used. From Eq. 3.3, it follows that

$$\arg w'(t_0) = \arg f'(z_0) + \theta \tag{3.6}$$

which implies that the transformation $w = f(z)$ rotates the direction of the tangent vector at z_0 through the angle of rotation, $\arg f'(z_0)$, resulting in the angle of inclination of the tangent vector at w_0. Figure 3.2 shows these observations.

It should be noted that the angle $\arg f'(z_0)$ exists if $f'(z_0) \neq 0$. When $f'(z_0) = 0$, $\arg f'(z_0)$ is indeterminate and Eq. 3.6 is invalid. Such a point z_0 is called a critical point of f. For the complex potential $\Omega(z)$, the points where the derivative $\Omega'(z)$ are zero are the stagnation points. Thus, the critical points in the mapping function $\Omega(z)$ are the stagnation points.

Example 3.1 Let us consider the half line C: $y = x$ ($y \geq 0$) and the transformation

$$w = f(z) = z^2 = x^2 - y^2 + 2xyi$$

The half line C is mapped onto C' given by

$$w = y^2 - y^2 + 2yyi = 2y^2 i$$

which results in $v = 2y^2$, the upper half of the v axis. The point $z_0 = 1 + i$ on the half line C is mapped onto $w_0 = 2i$ on C', as shown in Fig. 3.3.

At the point z_0, the angle of the tangent vector is $\theta = \pi/4$. At the point w_0, the angle of the tangent vector is $\arg w' = \pi/2$. Since $f'(z) = 2z$, it follows that $f'(z_0) = 2 + 2i$ and the argument of f' at z_0 is $\pi/4$, which confirms that Eq. 3.6 is

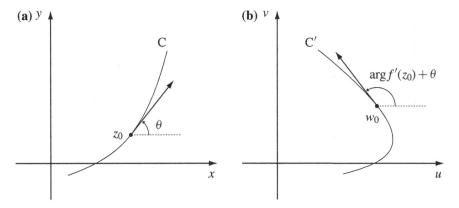

Fig. 3.2 Curves under a mapping $w = f(z)$. **a** Tangent vector at z_0 making an angle θ with the x direction. **b** Tangent vector at w_0 making an angle $\arg f'(z_0) + \theta$ with the u direction

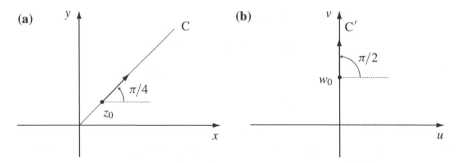

Fig. 3.3 Curves under a mapping $w = z^2$. **a** Tangent vector at z_0 making an angle $\pi/4$ with the x direction. **b** Tangent vector at w_0 making an angle $\pi/2$ with the u direction

satisfied. The point $z = 0$ is the critical point of the transformation $w = z^2$, where $\arg f'(z)$ cannot be defined.

Suppose that C_1 and C_2 are the two curves passing through the point z_0 which are respectively mapped onto the image curves C_1' and C_2' passing through the point w_0. If the angle at z_0 between C_1 and C_2 is equal to the angle at w_0 between C_1' and C_2' both in magnitude and sense, the mapping is said to be conformal. In Fig. 3.4, for instance, the angle δ from C_1 to C_2 in the z plane is the same in magnitude and sense as the angle from C_1' to C_2' in the w plane, and thus, the mapping is conformal at z_0.

Theorem 3.1 (Conformal mapping) *If $f(z)$ is analytic, the mapping $w = f(z)$ is conformal except at critical points where the derivative $f'(z)$ is zero.*

Proof Let C_1 and C_2 be the two curves passing through the point z_0 in the z plane and θ_1 and θ_2 be the angles of the vectors tangent to C_1 and C_2 at z_0, respectively, as shown in Fig. 3.5a. The angle at z_0 between the curves C_1 and C_2 is $\theta_2 - \theta_1$. Under

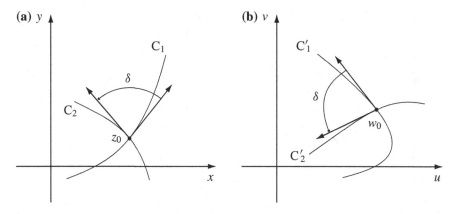

Fig. 3.4 Conformal mapping. **a** Angle δ from C_1 to C_2 in the z plane. **b** Angle δ from C_1' to C_2' in the w plane

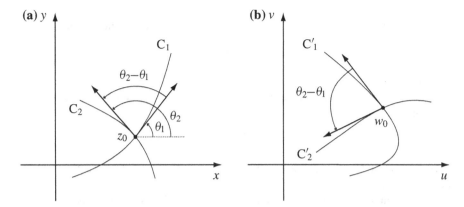

Fig. 3.5 Two curves under a mapping $w = f(z)$. **a** Tangent vectors at z_0 with an included angle $\theta_2 - \theta_1$. **b** Tangent vectors at w_0 with an included angle $\theta_2 - \theta_1$

the mapping $w = f(z)$, the curves C_1 and C_2 are respectively mapped onto the image curves C_1' and C_2' in the w plane, as shown in Fig. 3.5b.

Since $f(z)$ is analytic, its derivative $f'(z)$ exists and $\arg f'(z)$ can be defined as long as $f'(z) \neq 0$. From Eq. 3.6, the angles of the vectors tangent to the image curves C_1' and C_2' at w_0 are given by $\arg f'(z_0) + \theta_1$ and $\arg f'(z_0) + \theta_2$, respectively. Hence, the angle at w_0 between the curves C_1' and C_2' is $\theta_2 - \theta_1$, as shown in Fig. 3.5b. The angle between the two curves is preserved, both in magnitude and sense; thus, the mapping $w = f(z)$ is conformal except at critical points where $f'(z) = 0$. □

Example 3.2 Let us consider the half lines $C_1: y = x$ $(y \geq 0)$ and $C_2: x = 1$ $(y \geq 0)$, which intersect at $z_0 = 1+i$, as shown in Fig. 3.6a. The angle between the two curves at z_0 is $\pi/4$. As discussed in Example 3.1, the transformation $w = f(z) = z^2$ maps C_1 onto the upper half of the v axis and the point z_0 onto $w_0 = 2i$. The half line C_2 is mapped onto the curve C_2' given by

$$w = (1 + iy)^2 = 1 - y^2 + 2yi$$

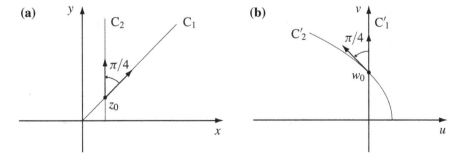

Fig. 3.6 Two curves under a mapping $w = z^2$. **a** Tangent vectors at z_0, where $f'(z_0) \neq 0$, with an included angle $\pi/4$. **b** Tangent vectors at w_0 with an included angle $\pi/4$

which results in $u = 1 - y^2$ and $v = 2y$, that is, the upper half of the parabola

$$v^2 = -4(u - 1)$$

as shown in Fig. 3.6b.

From the parametric equation of C'_2 with y as the parameter, the vector tangent to C'_2 is obtained as

$$w'(y) = \frac{dw}{dy} = -2y + 2i$$

At $w_0 = w(1)$, $w'(1) = -2 + 2i$ and it follows that the angle of the vector tangent to C'_2 is $\arg w'(1) = (3/4)\pi$.

At w_0, the angle of the vector tangent to C'_1 is $\pi/2$, and thus, the angle between the image curves C'_1 and C'_2 is $\pi/4$, which is equal to the angle between the two curves at z_0 in the z plane. The conformality of the transformation $w = z^2$ at $z_0 = 1 + i$ is confirmed.

Example 3.3 Let us consider the half lines $C_1: y = x$ ($y \geq 0$) and $C_2: x = 0$ ($y \geq 0$), which intersect at $z_0 = 0$, as shown in Fig. 3.7a. The angle between the two curves at z_0 is $\pi/4$. As discussed in Example 3.1, the transformation $w = f(z) = z^2$ maps C_1 onto the upper half of the v axis. The point z_0 is mapped onto $w_0 = 0$. The half line C_2 is mapped onto the curve C'_2 given by

$$w = (iy)^2 = -y^2$$

which results in $u = -y^2$, the negative u half-axis, as shown in Fig. 3.7b.

At w_0, the angle of the vector tangent to C'_1 is $\pi/2$, and that to C'_2 is π. Hence, the angle between the image curves C'_1 and C'_2 is $\pi/2$, which is twice the angle between the two curves C_1 and C_2 at z_0 in the z plane. The point $z = 0$ is the critical point of the transformation $w = z^2$, where the conformality is not satisfied.

Theorem 3.2 (Angle magnification at a critical point) *If $f(z)$ is analytic at z_0 and if $f'(z_0) = \cdots = f^{(m-1)}(z_0) = 0$ and $f^{(m)}(z_0) \neq 0$, then the angle between any two curves passing through z_0 is multiplied by m under the transformation $w = f(z)$.*

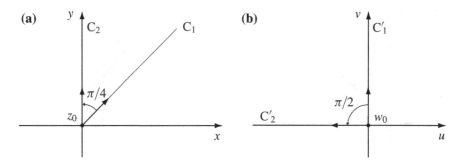

Fig. 3.7 Two curves under a mapping $w = z^2$. **a** Tangent vectors at z_0, where $f'(z_0) = 0$, with an included angle $\pi/4$. **b** Tangent vectors at w_0 with an included angle $\pi/2$

Proof The Taylor series expansion[1] for f at z_0 is given by

$$f(z) = f(z_0) + \alpha_m(z - z_0)^m + \alpha_{m+1}(z - z_0)^{m+1} + \alpha_{m+2}(z - z_0)^{m+2} + \cdots$$

where $\alpha_n = f^{(n)}(z_0)/n!$ and $\alpha_1 = \cdots = \alpha_{m-1} = 0$ is used. This can be rewritten as

$$f(z) - f(z_0) = (z - z_0)^m[\alpha_m + \alpha_{m+1}(z - z_0) + \alpha_{m+2}(z - z_0)^2 + \cdots]$$
$$= (z - z_0)^m g(z) \tag{3.7}$$

where

$$g(z) = \alpha_m + \alpha_{m+1}(z - z_0) + \alpha_{m+2}(z - z_0)^2 + \cdots$$

which is analytic at z_0 and

$$g(z_0) = \alpha_m \neq 0 \tag{3.8}$$

From Eq. 3.7, it follows that

$$\arg(w - w_0) = \arg[f(z) - f(z_0)]$$
$$= m \arg(z - z_0) + \arg g(z) \tag{3.9}$$

where the identity Eq. 2.34 is used.

For a curve C passing through z_0, the angle θ of the tangent vector at z_0 is given by

$$\theta = \lim_{z \to z_0} \arg(z - z_0) \tag{3.10}$$

Hence, the angle ϕ of the vector tangent to the image curve C' at w_0 is

$$\phi = \lim_{w \to w_0} \arg(w - w_0)$$
$$= \lim_{z \to z_0} (m \arg(z - z_0) + \arg g(z))$$
$$= m \lim_{z \to z_0} \arg(z - z_0) + \arg g(z_0)$$
$$= m\theta + \arg \alpha_m \tag{3.11}$$

where Eqs. 3.8–3.10 are used.

Let C_1 and C_2 be the two curves passing through the point z_0 which are respectively mapped onto the image curves C'_1 and C'_2 passing through the point w_0. From Eq. 3.11, the angles of the tangent vectors to C'_1 and C'_2, ϕ_1 and ϕ_2, are related to the angles of the tangent vectors to C_1 and C_2, θ_1 and θ_2, as

$$\phi_2 - \phi_1 = m(\theta_2 - \theta_1)$$

---

[1] The existence of Taylor series for analytic functions is proved in Chap. 5.

which implies that the angle from C_1' to C_2' is m times as large as the angle from C_1 to C_2.   □

Example 3.4 Let us consider the transformation $w = f(z) = z^n$, where n is an integer ($n \geq 2$). Since $f'(z) = nz^{n-1}$ and $f'(0) = 0$, $z = 0$ is the critical point. Also, since $f'(0) = \cdots = f^{(n-1)}(0) = 0$ and $f^{(n)}(0) = n! \neq 0$, according to Theorem 3.2, the angle between any two curves passing through $z = 0$ is multiplied by n under the transformation $w = z^n$. For instance, when $n = 2$, the angle is doubled as seen in Example 3.3.

A mapping that preserves the magnitude of the angle between two curves but not necessarily the sense is called an isogonal mapping. A conformal mapping is an isogonal mapping that also preserves the sense of the angle.

Example 3.5 The transformation $w = \bar{z}$, a reflection in the real axis, is isogonal but not conformal. If $f(z)$ is conformal, $w = f(\bar{z})$ is isogonal but not conformal. This is because $\bar{z}$ is not analytic.

3.1.2 General Transformations

General transformations; translation, rotation, stretching, and inversion are summarized.

3.1.2.1 Translation

Objects in the z plane are translated in the direction of vector β by the function

$$w = z + \beta \tag{3.12}$$

where β is the complex constant.

3.1.2.2 Rotation

Let $z = re^{i\theta}$, then the function

$$w = e^{i\phi}z \tag{3.13}$$

maps z onto

$$w = re^{i(\theta+\phi)} \tag{3.14}$$

Hence, objects in the z plane are rotated through an angle ϕ. If $\phi > 0$, the rotation is counterclockwise, while if $\phi < 0$, the rotation is clockwise.

3.1.2.3 Stretching

Objects in the z plane are stretched in the direction z by the function

$$w = az \tag{3.15}$$

if the real constant $a > 1$, while if $0 < a < 1$, they are contracted.

3.1.2.4 Inversion

Let $z = re^{i\theta}$ and $w = \rho e^{i\phi}$, then the function

$$w = \frac{1}{z} \tag{3.16}$$

maps z onto

$$w = \rho e^{i\phi} = \frac{1}{re^{i\theta}} = \frac{1}{r}e^{-i\theta} \tag{3.17}$$

which gives

$$\begin{cases} \rho = 1/r \\ \phi = -\theta \end{cases} \tag{3.18}$$

The unit circle $|z| = r = 1$ is mapped onto the unit circle $|w| = \rho = 1$.

Example 3.6 Figure 3.8 exemplifies the aforementioned general transformations. The transformation $w = z + (2 + 4i)$ translates the point $z = 1 + i$ to $w = 3 + 5i$, $w = e^{(\pi/2)i}z$ rotates the point $z = 1 + i = \sqrt{2}e^{(\pi/4)i}$ to $w = \sqrt{2}e^{(3\pi/4)i} = -1 + i$, and $w = 3z$ stretches the modulus of $z = 1 + i$, $(|z| = \sqrt{2})$ by the factor 3.

From Eq. 3.18, the transformation $w = 1/z$ can be considered as an inversion with respect to the unit circle $|z| = 1$ and a reflection in the real axis, as shown in Fig. 3.9. The points z exterior to the circle $|z| = 1$ are mapped onto the nonzero

Fig. 3.8 Translation, rotation, and stretching

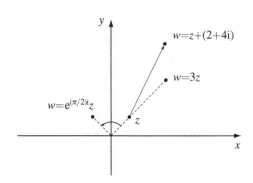

Fig. 3.9 Inversion

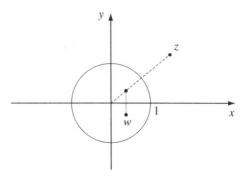

points interior to the circle $|w| = 1$, and conversely, the nonzero points interior to $|z| = 1$ are mapped onto the points exterior to $|w| = 1$.

Let a point $w = u + iv$ be the image of a nonzero point $z = x + iy$ under inversion $w = 1/z$, then

$$\begin{cases} x = \dfrac{u}{u^2 + v^2} \\[2mm] y = \dfrac{-v}{u^2 + v^2} \end{cases} \tag{3.19}$$

When the real numbers a, b, c, and d satisfy $b^2 + c^2 > 4ad$, the equation

$$a(x^2 + y^2) + bx + cy + d = 0 \tag{3.20}$$

represents an arbitrary circle or straight line, where $a \neq 0$ for a circle and $a = 0$ for a straight line. Substituting Eq. 3.19 into Eq. 3.20 yields

$$d(u^2 + v^2) + bu - cv + a = 0 \tag{3.21}$$

which also represents a circle or straight line.

If $a \neq 0$, Eq. 3.20 gives a circle in the z plane. For $d \neq 0$, a circle not passing through the origin is mapped onto a circle (since $d \neq 0$) not passing through the origin (since $a \neq 0$) in the w plane. For $d = 0$, a circle through the origin is mapped onto a line (since $d = 0$) not passing through the origin (since $a \neq 0$) in the w plane.

If $a = 0$, Eq. 3.20 gives a straight line in the z plane. For $d \neq 0$, a line not passing through the origin is mapped onto a circle (since $d \neq 0$) passing through the origin (since $a = 0$) in the w plane. For $d = 0$, a line through the origin is mapped onto a line (since $d = 0$) passing through the origin (since $a = 0$) in the w plane.

Therefore, the transformation $w = 1/z$ maps every straight line or circle onto a circle or straight line, as shown in Fig. 3.10.

Example 3.7 According to Eqs. 3.20 and 3.21, a vertical line $x = 1/3$ ($a = c = 0$, $b = -3$, and $d = 1$) is mapped by $w = 1/z$ onto the image

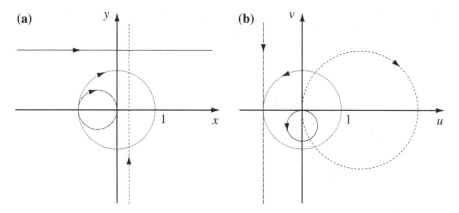

Fig. 3.10 Mapping by $w = 1/z$. **a** Straight lines or circles in the z plane. **b** Circles or straight lines in the w plane

$$u^2 + v^2 - 3u = \left(u - \frac{3}{2}\right)^2 + v^2 - \frac{9}{4} = 0$$

which is a circle of radius $3/2$ with its center at $3/2$. The circle passes through the origin.

3.2 Mapping by Elementary Functions

This section further explores how various curves and regions are mapped by elementary functions. Applications to flow problems are included.

3.2.1 Linear Function

The transformation given by the linear function

$$w = \alpha z + \beta \tag{3.22}$$

where α and β are the complex constants, is called the linear transformation. Equation 3.22 can be written in terms of the successive transformations

$$\begin{cases} w = \zeta + \beta \\ \zeta = e^{i\theta} \tau \\ \tau = az \end{cases} \tag{3.23}$$

where $\alpha = ae^{i\theta}$. Hence, a linear transformation can be considered as a combination of the transformations of translation, rotation, and stretching.

Example 3.8 Let us find the complex potential for uniform flow with a flow velocity q_u/A in the direction making an angle δ with the x direction. The x and y components of the velocity are

$$\begin{cases} V_x = \dfrac{q_u}{A} \cos \delta \\[2mm] V_y = \dfrac{q_u}{A} \sin \delta \end{cases}$$

and the complex velocity W becomes

$$W = -\frac{d\Omega}{dz} = V_x - iV_y = \frac{q_u}{A}(\cos \delta - i \sin \delta) = \frac{q_u}{A}e^{-i\delta}$$

Then, integration gives

$$\Omega(z) = -\frac{q_u}{A}e^{-i\delta}z \tag{3.24}$$

where the constant of integration is omitted. The resultant equipotential lines and streamlines are shown in Fig. 3.11a.

The equipotential lines and streamlines are the level curves, defined by $\Phi = p$ and $\Psi = s$ for various values of p and s, respectively. In the Ω plane, these lines are respectively parallel to the Ψ and Φ axes, as shown in Fig. 3.11b. This implies that the complex potential given by Eq. 3.24 maps uniform inclined flow in the z plane onto uniform flow parallel to the Φ axis in the Ω plane by rotating the flow profiles in the z plane through an angle $-\delta$.

In general, the function $\Omega = \Omega(z)$ is the mapping of arbitrary flow defined by the complex potential $\Omega(z)$ in the z plane onto uniform flow parallel to the Φ axis in

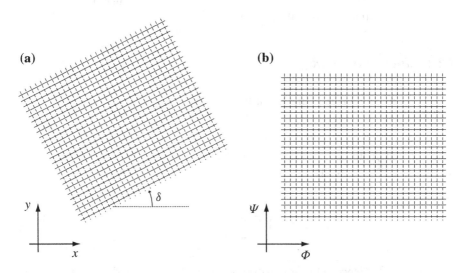

(a)   **(b)**

Fig. 3.11 Mapping by $\Omega = \Omega(z)$. **a** Uniform inclined flow in the z plane. **b** Complex potential in the Ω plane

the Ω plane. Since the complex potential $\Omega(z)$ is analytic, the mapping is conformal except at critical points.

3.2.2 Power Function

The transformation given by the power function

$$w = z^n \tag{3.25}$$

is conformal, except at the critical point $z = 0$, where $w' = nz^{n-1} = 0$. This mapping is most easily described in terms of polar coordinates. Let $z = re^{i\theta}$ and $w = \rho e^{i\phi}$, then

$$\rho e^{i\phi} = r^n e^{in\theta} \tag{3.26}$$

which gives

$$\begin{cases} \rho = r^n \\ \phi = n\theta \end{cases} \tag{3.27}$$

The angle at the critical point $z = 0$ is multiplied by n under the mapping, as asserted by Theorem 3.2 and discussed in Example 3.4. Hence, the sector $0 \leq \theta \leq \pi/n$ is mapped by z^n onto the upper half plane, $\rho \geq 0$ and $0 \leq \phi \leq \pi$, in the w plane, as shown in Fig. 3.12. Although the power function is most definitely not conformal at $z = 0$, which is the critical point, it must be emphasized that the mapping preserves angles everywhere else.

Example 3.9 Let us consider flow around a right-angle ($\pi/2$) corner, as shown in Fig. 3.13. A fluid enters with a flow velocity $q_u/A = 2$ through the first quadrant ($x > 0, y > 0$) in the negative x direction and is forced to turn a corner near the origin to leave upward parallel to the y axis. The x and y axes form impermeable boundaries.

With the power function $\Omega = az^2$, the angle at the critical point $z = 0$ in the z plane is doubled in the Ω plane, and thus, the right-angle sector is mapped onto the

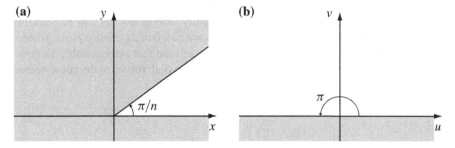

Fig. 3.12 Mapping by $w = z^n$. **a** Sector $0 \leq \theta \leq \pi/n$ in the z plane. **b** Upper half plane $0 \leq \phi \leq \pi$ in the w plane

Fig. 3.13 Flow around a
right-angle corner

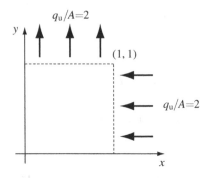

upper half plane. While the positive x half-axis is mapped onto the positive Φ axis,
the positive y half-axis is mapped onto the negative Φ axis. The Φ axis has $\Psi = 0$
(constant) and forms an impermeable boundary, which corresponds to the impermeable boundaries of the x and y axes in the z plane. This implies that flow around a
right-angle corner in the z plane can be modeled by the mapping $\Omega(z) = az^2$.

The complex potential $\Omega(z) = az^2$ gives

$$\begin{cases} \Phi = a(x^2 - y^2) \\ \Psi = 2axy \end{cases} \tag{3.28}$$

To express the flow velocity $q_u/A = 2$ between $z = (1, 0)$ and $z = (1, 1)$, let
$\Psi(1, 0) = 0$ and $\Psi(1, 1) = 2$. From Eq. 3.28, it follows that $a = 1$, and thus, the
flow around a corner with a flow velocity $q_u/A = 2$ is given by

$$\Omega(z) = z^2 \tag{3.29}$$

Figure 3.14 shows the corresponding equipotential lines and streamlines.

The complex velocity is given by

$$W = -\frac{d\Omega}{dz} = -2z$$

and from the solution of $W = 0$, a stagnation point is found at $z = 0$, where the
equipotential line and streamline are not mutually orthogonal, as can be seen in
Fig. 3.14. From the viewpoint of mapping, the point $z = 0$ is the critical point, where
the conformality is not satisfied. It is interesting to note that orthogonality in flow
properties and conformality in mapping properties are both ruined at the point where
$d\Omega/dz = 0$.

(a)   **(b)**

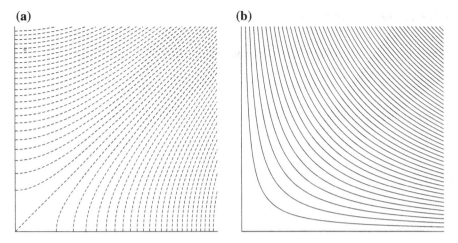

Fig. 3.14 Flow around a right-angle corner with $\Omega(z) = z^2$. **a** Equipotential lines. **b** Streamlines

3.2.3 Bilinear Function

The transformation given by the bilinear function

$$w = \frac{\alpha z + \beta}{\gamma z + \delta} \quad (\alpha\delta - \beta\gamma \neq 0) \tag{3.30}$$

where α, β, γ, and δ are the complex constants, is called the bilinear transformation, Möbius transformation, or linear-fractional transformation. The condition ensures that the transformation excludes the case of $w = $ constant and that the bilinear function is conformal by satisfying

$$w' = \frac{\alpha\delta - \beta\gamma}{(\gamma z + \delta)^2} \neq 0 \tag{3.31}$$

Special cases of bilinear functions include

$$\begin{cases} w = z + \beta & \text{(Translation)} \\ w = \alpha z & \text{(Stretching with } |\alpha| \neq 1) \\ w = \alpha z = e^{i\phi}z & \text{(Rotation with } |\alpha| = 1) \\ w = 1/z & \text{(Inversion)} \end{cases} \tag{3.32}$$

which implies that Eq. 3.30 has all the features of general transformations discussed in Sect. 3.1.2. From a practical point of view, bilinear functions contain many useful properties, which are further discussed in Sect. 3.4 with practical examples.

3.2.4 Exponential Function

The transformation given by the exponential function

$$w = e^z \tag{3.33}$$

is conformal in the entire complex plane. Let $z = x + iy$ and $w = \rho e^{i\phi}$, then

$$\rho e^{i\phi} = e^x e^{iy} \tag{3.34}$$

which gives

$$\begin{cases} \rho = |w| = e^x \\ \phi = \arg w = y \end{cases} \tag{3.35}$$

Under the mapping, a horizontal line $y = y_0$, as x varies $-\infty$ to ∞, is mapped onto the ray emanating from the origin which makes an angle $\arg w = y_0$ with the u axis (solid lines in Fig. 3.15). It should be noted that since $\arg w = y_0 = y_0 + 2n\pi$, where n is any integer, horizontal lines $y = y_0 + 2n\pi$ are all mapped onto the same ray.

A vertical line $x = x_0$ is mapped onto the circle $|w| = e^{x_0}$ (dashed lines in Fig. 3.15). The image moves counterclockwise around the circle. The mapping is many-to-one; each point on the circle $|w| = e^{x_0}$ is the image of an infinite number of points, spaced 2π units apart, along the line $x = x_0$.

Figure 3.16 shows the mapping of a rectangle $-x_0 \le x \le x_0$ and $-y_0 \le y \le y_0$ in the z plane onto a region bounded by concentric circles and rays in the w plane. The mapping is one-to-one if $2y_0 < 2\pi$. In particular, if $y_0 = \pi$, the rectangular region is mapped, in a one-to-one manner, onto the annulus which has a branch cut along the negative u half-axis.

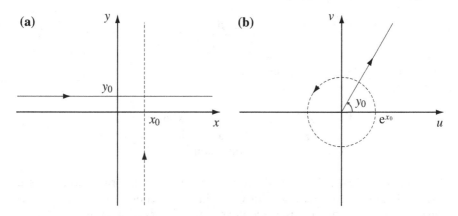

Fig. 3.15 Mapping by $w = e^z$. **a** Horizontal and vertical lines in the z plane. **b** Ray and circle in the w plane

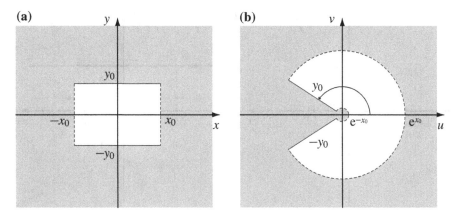

Fig. 3.16 Mapping by $w = e^z$. **a** Rectangular region in the z plane. **b** Fan-shaped region in the w plane

3.2.5 Logarithmic Function

In general, the mapping by the inverse $z = f^{-1}(w)$ of $w = f(z)$ is obtained by interchanging the roles of the z plane and the w plane in the mapping by $w = f(z)$. Hence, the transformation given by the logarithmic function

$$w = \ln z \tag{3.36}$$

is obtained by interchanging the roles of z and w in the mapping by $w = e^z$, considered in the previous section.

Since the mapping $w = e^z$ is many-to-one, in reverse, the mapping $w = \ln z$ is multi-valued (one-to-many). A single-valued branch of the function can be specified by making a branch cut, as shown by a thick line in Fig. 3.17a.

The image of a point $z = re^{i\theta}$ moving outward from the origin along the ray θ (a solid line in Fig. 3.17a) is the point whose coordinates in the w plane are $(u, v) = (\ln r, \theta)$, that is, the entire length of the horizontal line $v = \theta$ (a solid line in Fig. 3.17b). These lines fill the strip $-\pi < v \leq \pi$ as θ varies between $-\pi$ and π (dashed lines in Fig. 3.17). Hence, the logarithmic function maps the plane with the branch cut onto the region $-\pi < v \leq \pi$, in a one-to-one manner, as shown in Fig. 3.17b.

The mapping $w = \ln z + 2\pi i$ differs from $w = \ln z$ by the translation term $2\pi i$, and thus, this function maps the z plane with the branch cut onto the strip $\pi < v \leq 3\pi$. The same is true for each of the mappings $w = \ln z + 2n\pi i$, where n is an integer, and the infinitely many strips of width 2π cover the whole w plane without overlapping. The collection of infinitely many strips is called a Riemann surface corresponding to the function $w = \ln z$. The concept of Riemann surface is of use to obtain the various values of multi-valued function in a continuous manner.

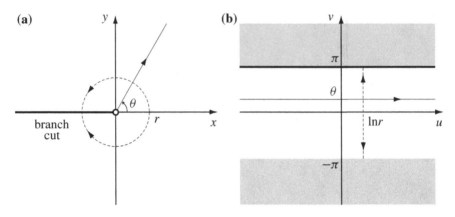

Fig. 3.17 Mapping by $w = \ln z$ with a branch cut. **a** Ray and circle in the z plane. **b** Horizontal and vertical lines in the w plane

Example 3.10 The complex potential for a source or sink at the origin is given by

$$\Omega(z) = -\frac{q_{\mathrm{w}}}{2\pi h} \ln z \tag{3.37}$$

the flow profile of which is as shown in Fig. 3.18a.

Since a logarithmic function $\ln z$ is not analytic at $z = 0$, the location of a source or sink is the singular point. An actual well (source or sink), however, has a finite wellbore radius, r_{w}. The wellbore forms an internal boundary, which excludes the

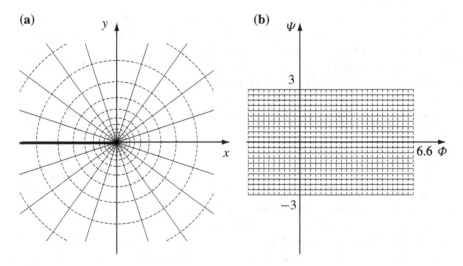

Fig. 3.18 Mapping by $\Omega = -(3/\pi) \ln z$. **a** Source or sink in the z plane. **b** Complex potential in the Ω plane

region $|z| < r_w$ from the domain of interest; thus, the singularity problem of Eq. 3.37, which occurs if $z = 0$, can be avoided.

For $q_w/h = 6$ and $r_w = 0.001$, for instance, Eq. 3.37 yields

$$\begin{cases} \Phi = -(3/\pi)\ln r \\ \Psi = -(3/\pi)\theta \end{cases}$$

the ranges of which are

$$\begin{cases} -\infty < \Phi \leq -(3/\pi)\ln 0.001 \simeq 6.6 \\ -3 \leq \Psi < 3 \end{cases}$$

The mapping of the equipotential lines and streamlines in the z plane onto the Ω plane is obtained as shown in Fig. 3.18b.

3.3 Applications of Conformal Mapping

With the aid of composition and the Joukowski transformation, Solution Strategy is accomplished to solve Motivating Problem 4.

3.3.1 Composition

A useful hallmark of conformal mapping is that the composition of multiple conformal mappings is also conformal. Thus, a complicated mapping can be constructed by applying a sequence of elementary conformal mappings. This relies on the fact that the composition of analytic functions is also analytic.

Let $w = f(\chi)$ be an analytic function of the complex variable $\chi = \xi + i\eta$ and $\chi = g(z)$ be an analytic function of the complex variable $z = x + iy$; then the composition $w = f(g(z))$ is an analytic function, as shown by the chain rule

$$\frac{dw}{dz} = \frac{dw}{d\chi}\frac{d\chi}{dz} \tag{3.38}$$

where the range of the function $\chi = g(z)$ is contained in the domain of definition of the function $w = f(\chi)$.

The conformality of the composition of two conformal mappings can be confirmed by the fact from the chain rule that $dw/dz \neq 0$ provided $dw/d\chi \neq 0$ and $d\chi/dz \neq 0$. If both f and g are one-to-one, the composition w also yields a one-to-one mapping.

Example 3.11 In Sect. 2.5.3, the complex potential for a source or sink at z_w is derived as

$$\Omega(z) = -\frac{q_w}{2\pi h}\ln(z - z_w)$$

The same result can be obtained by the composite function $\Omega(\chi(z))$ of the two mappings

$$\begin{cases} \chi(z) = z - z_w \\ \Omega(\chi) = -\dfrac{q_w}{2\pi h} \ln \chi \end{cases}$$

where the first equation is the translation of z_w to the origin (Eq. 3.12) and the second equation is for a source or sink at the origin (Eq. 2.110).

Example 3.12 Suppose that uniform flow over the circular pillar of radius $R = 1$, considered in Motivating Problem 3, is now making an angle $\delta = \pi/6$ with the x axis. The complex potential can be obtained as the composite function $\Omega(\chi(z))$ of the two mappings

$$\begin{cases} \chi(z) = e^{-i\delta} z \\ \Omega(\chi) = -\dfrac{q_u}{A} \left(\chi + \dfrac{R^2}{\chi} \right) \end{cases}$$

where the first equation is the clockwise rotation through the angle δ (Eq. 3.13) and the second equation is for uniform flow in the ξ direction over a circular object of radius R (Eq. 2.153). Figure 3.19 shows the two mappings and it follows that

$$\Omega(z) = -\frac{q_u}{A} \left(e^{-i\delta} z + \frac{e^{i\delta} R^2}{z} \right) \tag{3.39}$$

which is conformal except at the critical points.

For $q_u/A = 1$, $R = 1$, and $\delta = \pi/6$, the solution becomes

$$\Omega(z) = -\frac{\sqrt{3} - i}{2} z - \frac{\sqrt{3} + i}{2} \frac{1}{z}$$

and the equipotential lines and streamlines are obtained as shown in Fig. 3.20, where the flow profiles in Fig. 2.20 are rotated through an angle of $\delta = \pi/6$.

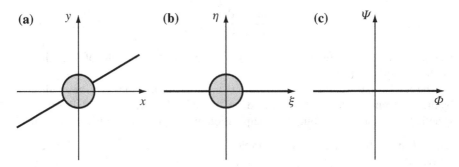

Fig. 3.19 Composition of conformal mappings. **a** Uniform inclined flow over a pillar in the z plane. **b** Uniform flow over a pillar in the χ plane. **c** Complex potential in the Ω plane

(a) **(b)**

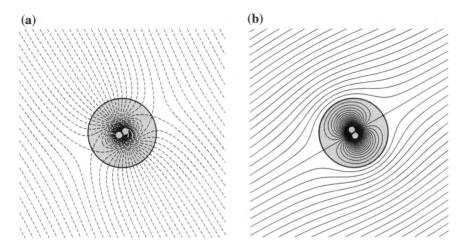

Fig. 3.20 Uniform inclined flow over a pillar. **a** Equipotential lines. **b** Streamlines

The complex velocity is given by

$$W = -\frac{d\Omega}{dz} = \frac{\sqrt{3} - i}{2} - \frac{\sqrt{3} + i}{2}\frac{1}{z^2}$$

and from the solution of $W = 0$ it follows that $z^2 = 1/2 + (\sqrt{3}/2)i = e^{(\pi/3)i}$. Hence, stagnation points are found at $z = 2^{(\pi/6)i} = \cos \pi/6 + i \sin \pi/6 = \sqrt{3}/2 + i/2$ and $z = \cos 7\pi/6 + i \sin 7\pi/6 = -\sqrt{3}/2 - i/2$, where the orthogonality between equipotential lines and streamlines is ruined, as seen in Fig. 3.20.

Example 3.13 Let us consider flow around a cylindrical corner. A fluid enters with a flow velocity $q_u/A = 2$ into the first quadrant ($x > 0$, $y > 0$) in the negative x direction and is forced to turn a cylindrical corner of radius 0.5 to leave upward parallel to the y axis, as shown in Fig. 3.21. The x and y axes and the cylindrical corner form impermeable boundaries.

Fig. 3.21 Flow around a cylindrical corner

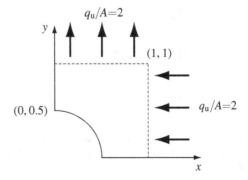

The complex potential can be obtained as the composite function $\Omega(\chi(z))$ of the two mappings

$$\begin{cases} \chi(z) = z^2 \\ \Omega(\chi) = a\left(\chi + \dfrac{0.25^2}{\chi}\right) \end{cases}$$

where the first equation is the transformation of the first quadrant in the z plane onto the upper half of the χ plane (Eq. 3.29), which transforms the cylinder of radius 0.5 onto the cylinder of radius 0.25. The second equation is for uniform flow in the negative ξ direction over a circular object of radius 0.25 (Eq. 2.153). Figure 3.22 shows the two mappings and it follows that

$$\Omega(z) = a\left(z^2 + \frac{0.25^2}{z^2}\right) \tag{3.40}$$

which is conformal except at the critical points.

The complex potential Eq. 3.40 gives

$$\begin{cases} \Phi = a(x^2 - y^2)\left[1 + \dfrac{0.25^2}{(x^2 + y^2)^2}\right] \\ \Psi = 2axy\left[1 - \dfrac{0.25^2}{(x^2 + y^2)^2}\right] \end{cases} \tag{3.41}$$

To express the flow velocity $q_u/A = 2$ between $z = (1, 0)$ and $z = (1, 1)$, let $\Psi(1, 0) = 0$ and $\Psi(1, 1) = 2$. From Eq. 3.41, it follows that $a = 1/0.984375$. The equipotential lines and streamlines are obtained as shown in Fig. 3.23.

The complex velocity is given by

$$W = -\frac{d\Omega}{dz} = -a\left(2z - \frac{0.125}{z^3}\right)$$

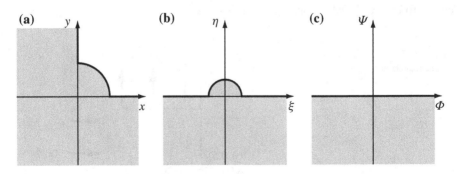

Fig. 3.22 Composition of conformal mappings. **a** Flow around a cylindrical corner in the z plane. **b** Upper half of uniform flow over a circular object in the χ plane. **c** Complex potential in the Ω plane

(a) **(b)**

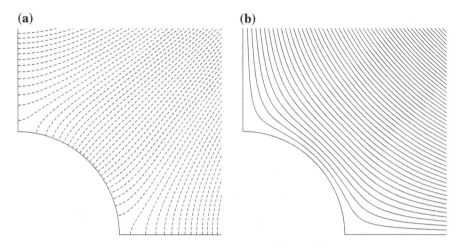

Fig. 3.23 Flow around a cylindrical corner. **a** Equipotential lines. **b** Streamlines

and from the solution of $W = 0$ it follows that $z^4 = 0.0625$. Hence, stagnation points are found at $z = \pm 0.5$ and $z = \pm 0.5i$, two of which are seen in Fig. 3.23, where the orthogonality between equipotential lines and streamlines is ruined.

3.3.2 Joukowski Transformation

The Joukowski transformation is given by a rational function

$$w = \frac{z^2 + R^2}{z} = z + \frac{R^2}{z} \tag{3.42}$$

where R is the nonzero real constant. The derivative of the function is

$$\frac{dw}{dz} = 1 - \frac{R^2}{z^2} = \frac{(z - R)(z + R)}{z^2} \tag{3.43}$$

which is equal to 0 if and only if $z = \pm R$. Hence, the mapping is conformal except at the critical points $z = \pm R$ and the singular point $z = 0$, where the function is not defined.

Let $z = re^{i\theta}$, then the Joukowski transformation is written as

$$w = re^{i\theta} + \frac{R^2}{r}e^{-i\theta}$$

$$= r(\cos\theta + i\sin\theta) + \frac{R^2}{r}(\cos\theta - i\sin\theta) \tag{3.44}$$

which gives

$$
\begin{cases}
u = \left(r + \dfrac{R^2}{r}\right)\cos\theta \\[3mm]
v = \left(r - \dfrac{R^2}{r}\right)\sin\theta
\end{cases}
\tag{3.45}
$$

For a circle of radius R in the z plane, it follows that

$$
\begin{cases}
u = 2R\cos\theta \\
v = 0
\end{cases}
\tag{3.46}
$$

the point of which lies on the real axis in the w plane, with $-2R \leq w \leq 2R$. Hence, the transformation squashes the circle of radius R, $|z| = R$, down to the real line segment $[-2R, 2R]$, as shown in Fig. 3.24.

For circles $|z| = r \neq R$, Eq. 3.45 reduces to

$$
\left(\frac{u}{r + R^2/r}\right)^2 + \left(\frac{v}{r - R^2/r}\right)^2 = 1
\tag{3.47}
$$

Hence, the Joukowski transformation maps concentric circles $|z| = r \neq R$ onto confocal ellipses with semi-axes $r \pm R^2/r$ and foci at $\pm 2R$ on the u axis. Examples for $|z| = 2R$ and $|z| = 3R$ are shown in Fig. 3.24.

The images of points outside the circle of radius R, $|z| > R$, fill the w plane except for the line segment $[-2R, 2R]$, as do the images of the nonzero points inside the circle of radius R, $|z| < R$. This is because of the reciprocal property of the Joukowski transformation $w(z) = w(R^2/z)$. Two points z and R^2/z in the z plane are mapped onto the same image point in the w plane, and thus, the mapping is two-to-one.

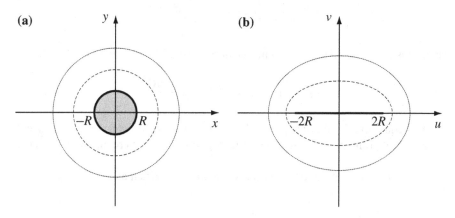

Fig. 3.24 Mapping by the Joukowski transformation. **a** Circles of $|z| = R, 2R,$ and $3R$ in the z plane. **b** Line segment $[-2R, 2R]$ and confocal ellipses in the w plane

The inverse mapping is obtained by solving Eq. 3.42 for z as

$$z = \frac{w + (w^2 - 4R^2)^{1/2}}{2} \tag{3.48}$$

which is a one-to-two mapping, since the term $(w^2 - 4R^2)^{1/2}$ has two different roots. Every w except $\pm 2R$ corresponds to two different points z; one lies outside the circle $|z| = R$ and the other lies inside the circle.

Now, let us express $w \pm 2R$ in exponential form

$$\begin{cases} w - 2R = r_1 e^{i\theta_1} \\ w + 2R = r_2 e^{i\theta_2} \end{cases} \tag{3.49}$$

then, a branch of the function $(w^2 - 4R^2)^{1/2}$ in Eq. 3.48 is obtained by choosing only the positive square root as

$$(w^2 - 4R^2)^{1/2} = \sqrt{r_1 r_2} e^{i(\theta_1 + \theta_2)/2} \tag{3.50}$$

where r_1, r_2, θ_1, and θ_2 are defined as shown in Fig. 3.25.

If the ranges for the angles are taken to be $-\pi < \theta_1 \leq \pi$ and $-\pi < \theta_2 \leq \pi$, the function $(w^2 - 4R^2)^{1/2}$ becomes discontinuous only when either $w = -2R$ or $w = 2R$ is encircled, and is continuous when both $w = -2R$ and $w = 2R$ are encircled. Hence, the line segment $[-2R, 2R]$ on the u axis is the branch cut, across which $(w^2 - 4R^2)^{1/2}$ is discontinuous.

By setting a branch cut of the line segment $[-2R, 2R]$ and choosing a branch given by Eq. 3.50, the inverse mapping Eq. 3.48 becomes a one-to-one mapping, which maps the w plane except for the branch cut onto the exterior of the circle $|z| = R$ in the z plane.

• Solution to Task 4-1

The complex potential for uniform flow, in the direction making an angle δ with the x direction, over a thin obstacle can be obtained as the composite function $\Omega(\chi(z))$ of the two mappings

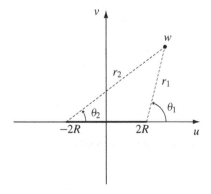

Fig. 3.25 Lengths r_1 and r_2 and angles θ_1 and θ_2 used for expressing $w - 2R$ and $w + 2R$

$$\begin{cases} \chi(z) = \dfrac{z + (z^2 - 4R^2)^{1/2}}{2} \\[2ex] \Omega(\chi) = -\dfrac{q_u}{A}\left(e^{-i\delta}\chi + \dfrac{e^{i\delta}R^2}{\chi}\right) \end{cases}$$

where the first equation is the inverse Joukowski transformation (Eq. 3.48) which maps the line segment $[-2R, 2R]$ onto the circle of radius R with its center at the origin and the second equation is for uniform inclined flow, in the direction making an angle of δ with the real axis, over the circular object of radius R (Eq. 3.39). The resultant complex potential is

$$\begin{aligned} \Omega(z) &= -\frac{q_u}{A}\left[\frac{e^{-i\delta}}{2}\left(z + (z^2 - 4R^2)^{1/2}\right) + \frac{2e^{i\delta}R^2}{z + (z^2 - 4R^2)^{1/2}}\right] \\[1ex] &= -\frac{q_u}{A}\left[\frac{e^{-i\delta}}{2}\left(z + (z^2 - 4R^2)^{1/2}\right) + \frac{e^{i\delta}}{2}\left(z - (z^2 - 4R^2)^{1/2}\right)\right] \\[1ex] &= -\frac{q_u}{A}\left(z\cos\delta - i(z^2 - 4R^2)^{1/2}\sin\delta\right) \end{aligned}$$

Substituting $q_u/A = 1$, $R = 1$, and $\delta = \pi/6$ yields

$$\Omega(z) = -\frac{\sqrt{3}}{2}z + \frac{i}{2}(z^2 - 4)^{1/2}$$

where the branch of $(z^2 - 4)^{1/2}$ is specified by Eq. 3.50. The thin shield $[-2, 2]$ forms the branch cut. Figure 3.26 shows the corresponding equipotential lines and streamlines.

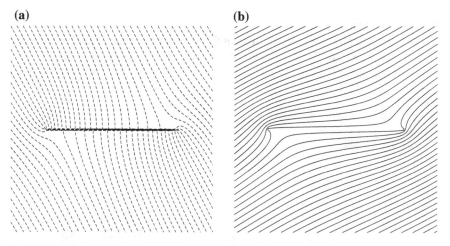

(a)   **(b)**

Fig. 3.26 Uniform inclined flow over a thin shield. **a** Equipotential lines. **b** Streamlines

The complex velocity is given by

$$W = -\frac{d\Omega}{dz} = \frac{\sqrt{3}}{2} - \frac{i}{2}\frac{z}{(z^2 - 4)^{1/2}} \tag{3.51}$$

and from the solution of $W = 0$, stagnation points are found at $z = \pm\sqrt{3}$ on the shield. Away from the thin shield, flow is essentially uniform as can be seen by evenly spaced inclined streamlines. When approaching the obstacle, flow is gradually distorted and the streamlines detour the shield and become denser near the edges of the shield, which implies higher velocities in the vicinities of the edges.

• **Solution to Task 4-2**
The complex velocity is given by Eq. 3.51, the real and imaginary parts of which give velocity components V_x and $-V_y$, respectively, and the absolute value of which gives the magnitude of velocity $|\mathbf{V}|$, as discussed in Sect. 2.5.5.

Figure 3.27 shows the velocity profiles along the x axis. Since the shield is the branch cut, different velocity profiles are obtained for the lower and upper sides of the shield. On either side, the magnitude of velocity $|\mathbf{V}|$ becomes infinite at the edges of the shield ($z = \pm 2$), and away from the shield, it approaches 1, the uniform flow velocity without an effect of the shield.

On the lower side of the shield (Fig. 3.27a), the velocity component V_x is negative between the left edge $x = -2$ and the stagnation point $x = -\sqrt{3}$. It increases to positive values toward the right edge $x = 2$ of the shield. Conversely, on the upper side (Fig. 3.27b), V_x is positive between the left edge and the other stagnation point $x = \sqrt{3}$ and decreases to negative values toward the right edge. Outside the shield, V_x is constant at $\sqrt{3}/2$, the uniform flow velocity in the x direction.

The velocity component V_y is zero along the shield between $x = -2$ and 2, since the shield is impermeable, across which no fluid can flow. At the edges of the shield,

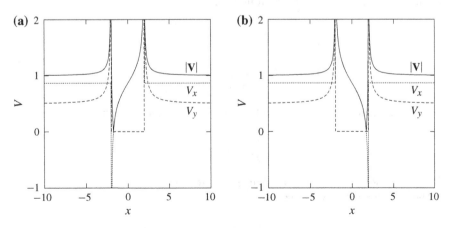

Fig. 3.27 Velocity profiles V_x, V_y, and $|\mathbf{V}|$ along the x axis for uniform inclined flow over a thin shield. **a** Lower side of the shield. **b** Upper side of the shield

Fig. 3.28 Velocity profiles V_x, V_y, and $|\mathbf{V}|$ along the y axis for uniform inclined flow over a thin shield

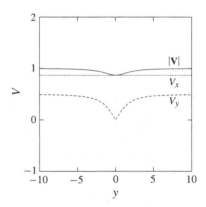

V_y abruptly increases to infinity and drastically decreases with the distance apart from the shield. At a substantial distance, V_y approaches $1/2$, the uniform flow velocity in the y direction.

Figure 3.28 shows the velocity profiles along the y axis, which are symmetric about the shield at $y = 0$. The magnitude of velocity $|\mathbf{V}|$ exhibits a slight reduction in the vicinity of the shield and approaches 1, the uniform flow velocity, away from the shield.

The velocity component V_x is constant at $\sqrt{3}/2$, the uniform flow velocity in the x direction. At the impermeable shield ($y = 0$), the velocity component V_y vanishes. Away from the shield, V_y gradually increases to $1/2$, which is the uniform flow velocity in the y direction.

• Solution to Task 4-3

Let us consider that the thin object is an infinite conductivity fracture rather than a shield. From Example 2.40, it is deduced that the desired complex potential is obtained as the composition of the following mappings:

$$\begin{cases} \chi(z) = \dfrac{z + (z^2 - 4R^2)^{1/2}}{2} \\[2mm] \Omega(\chi) = -\dfrac{q_u}{A}\left(e^{-i\delta}\chi - \dfrac{e^{i\delta}R^2}{\chi}\right) \end{cases} \tag{3.52}$$

The resultant complex potential, with $q_u/A = 1$, $R = 1$, and $\delta = \pi/6$, is

$$\Omega(z) = \frac{i}{2}z - \frac{\sqrt{3}}{2}(z^2 - 4)^{1/2}$$

Figure 3.29 shows the equipotential lines and streamlines.

The complex velocity is given by

$$W = -\frac{i}{2} + \frac{\sqrt{3}}{2}\frac{z}{(z^2 - 4)^{1/2}}$$

(a) **(b)**

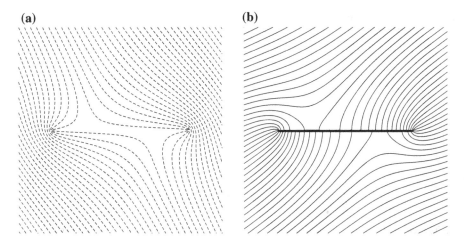

Fig. 3.29 Uniform inclined flow through an infinite conductivity fracture. **a** Equipotential lines. **b** Streamlines

and from the solution of $W = 0$, stagnation points are found at $z = \pm 1$ on the fracture. As is the case with the thin shield, the magnitude of velocity becomes infinite at the edges of the fracture ($z = \pm 2$).

Figure 3.30 shows the velocity profiles along the x axis. On the lower side of the fracture (Fig. 3.30a), the velocity component V_y is negative between the right edge $x = 2$ and the stagnation point $x = 1$. It increases to positive values toward the left edge $x = -2$ of the fracture. Conversely, on the upper side (Fig. 3.30b), V_y is positive between the right edge and the other stagnation point $x = -1$ and decreases to negative values toward the left edge. Outside the shield, V_y is constant at $1/2$, the uniform flow velocity in the y direction.

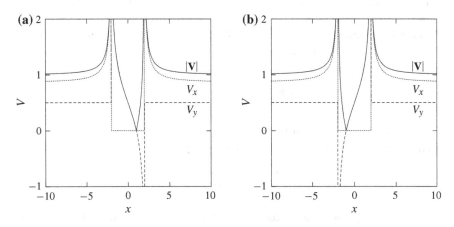

Fig. 3.30 Velocity profiles V_x, V_y, and $|V|$ along the x axis for uniform inclined flow through an infinite conductivity fracture. **a** Lower side of the fracture. **b** Upper side of the fracture

Fig. 3.31 Velocity profiles V_x, V_y, and $|\mathbf{V}|$ along the y axis for uniform inclined flow through an infinite conductivity fracture

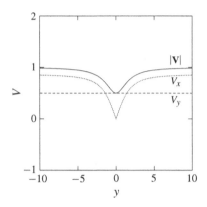

The velocity component V_x is zero along the fracture between $x = -2$ and 2, since the fracture is an equipotential line. At the edges of the fracture, V_x abruptly increases to infinity and drastically decreases with the distance apart from the fracture. At a substantial distance, V_x approaches $\sqrt{3}/2$, the uniform flow velocity in the x direction.

Figure 3.31 shows the velocity profiles along the y axis, which are symmetric about the fracture at $y = 0$. The magnitude of velocity $|\mathbf{V}|$ exhibits a reduction in the vicinity of the fracture and approaches 1, the uniform flow velocity, away from the fracture.

The velocity component V_y is constant at $1/2$, the uniform flow velocity in the y direction. At the infinite conductivity fracture ($y = 0$), the velocity component V_x vanishes. Away from the fracture, V_x gradually increases to $\sqrt{3}/2$, which is the uniform flow velocity in the x direction.

3.4 Möbius Transformation

As discussed in Sect. 3.2.3, Möbius transformations have all the features of general transformations: translation, rotation, stretching, and inversion. As revealed in this section, Möbius transformations map circles and lines onto other circles and lines, which is of great use in engineering problems.

3.4.1 Extended Complex Plane

To broaden the applicability of Möbius transformations, the extended complex plane is introduced. For a Möbius transformation given by

$$w = \frac{\alpha z + \beta}{\gamma z + \delta} \quad (\alpha\delta - \beta\gamma \neq 0) \tag{3.53}$$

there exists a unique w to each z for which $\gamma z + \delta \neq 0$. However, when $\gamma z + \delta = 0$ and $\gamma \neq 0$, that is, $z = -\delta/\gamma$, there is no w corresponding to this value of z, since the denominator vanishes.

To enlarge the domain of definition of the transformation, let $w = \infty$ be the image of $z = -\delta/\gamma$. The complex plane together with the point at infinity is called the extended complex plane. Also, the inverse mapping is obtained by solving for z in terms of w

$$z = \frac{\delta w - \beta}{-\gamma w + \alpha} \tag{3.54}$$

which is again a Möbius transformation. When $-\gamma w + \alpha = 0$ and $\gamma \neq 0$, let $w = \alpha/\gamma$ be the image of $z = \infty$. With these settings, the Möbius transformation becomes a one-to-one mapping of the extended z plane onto the extended w plane.

It thus follows that the Möbius transformation w maps the extended complex plane in a one-to-one manner onto itself with

$$\begin{cases} w(\infty) = \alpha/\gamma & \text{(if } \gamma \neq 0) \\ w(\infty) = \infty & \text{(if } \gamma = 0) \\ w(-\delta/\gamma) = \infty & \text{(if } \gamma \neq 0) \end{cases} \tag{3.55}$$

which makes the transformation w continuous on the extended z plane.

Example 3.14 Let us consider the Möbius transformation

$$w = \frac{\alpha z + \beta}{\gamma z + \delta} = \frac{z - i}{z + i} \tag{3.56}$$

which is known as the Cayley transformation. Since $\gamma = 1 \neq 0$, it follows that $w(\infty) = \alpha/\gamma = 1$ and $w(-\delta/\gamma) = w(-i) = \infty$.

For the unit circle in the w plane, $|w| = 1$, the corresponding points in the z plane satisfy

$$|z - i| = |z + i|$$

Squaring both sides gives

$$x^2 + (y - 1)^2 = x^2 + (y + 1)^2$$

which yields $y = 0$, the equation of the x axis.

The x axis divides the z plane into two portions, the upper and lower halves, and its image is the unit circle, $|w| = 1$, which divides the w plane into two portions, the interior and exterior of the circle. The image of the point $z = i$ is $w = 0$, and thus, the upper half of the z plane is mapped onto the interior of the unit circle in the w plane.

Figure 3.32 shows the mapping. The Cayley transformation maps the upper half plane Im $z > 0$ onto the open disk $|w| < 1$ and the boundary Im $z = 0$ onto the boundary $|w| = 1$ (thick lines in Fig. 3.32).

For instance, the horizontal line passing through $y = i$ (a dashed line in Fig. 3.32a), that is, $z = x + i$, is mapped as

$$w = u + iv = \frac{x}{x + 2i}$$

which gives

$$\begin{cases} xu - 2v - x = 0 \\ xv + 2u = 0 \end{cases}$$

Eliminating x yields

$$\left(u - \frac{1}{2}\right)^2 + v^2 = \frac{1}{4}$$

which is the circle of radius $1/2$ with its center at $w = 1/2$ (a dashed line in Fig. 3.32b).

When $\gamma = 0$, the condition $\alpha\delta - \beta\gamma \neq 0$ becomes $\alpha\delta \neq 0$, and the bilinear function reduces to a linear transformation. When $\gamma \neq 0$, Eq. 3.53 can be rewritten as

$$w = \frac{\alpha}{\gamma} + \frac{\beta\gamma - \alpha\delta}{\gamma} \frac{1}{\gamma z + \delta} \qquad (\alpha\delta - \beta\gamma \neq 0) \qquad (3.57)$$

Equation 3.57 can be written in terms of the successive transformations:

$$\begin{cases} w = \frac{\alpha}{\gamma} + \frac{\beta\gamma - \alpha\delta}{\gamma}\zeta \\ \zeta = 1/\tau \\ \tau = \gamma z + \delta \end{cases} \qquad (3.58)$$

Hence, a Möbius transformation can be considered as a linear transformation, followed by an inversion followed by another linear transformation.

It is clear that a linear transformation preserves shapes; lines onto lines and circles onto circles. As discussed in Sect. 3.1.2, an inversion maps every straight line or circle onto a circle or straight line, and thus, every Möbius transformation maps the class

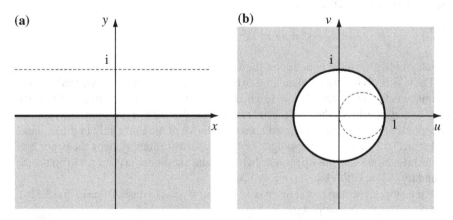

Fig. 3.32 Mapping by the Cayley transformation. **a** Upper half plane and $z = x + i$ in the z plane. **b** Unit disk and a circle passing through the origin in the w plane

of straight lines and circles onto itself. A half plane can be considered to be a family of parallel lines and a disk as a family of circles. Hence, a Möbius transformation maps the class of half planes and disks onto itself, as exemplified in Example 3.14.

3.4.2 Fixed Points

A fixed point of a transformation f is a point that is mapped onto itself, such that

$$w = f(z) = z \tag{3.59}$$

The identity mapping $w = z$, thus, has every point as a fixed point. For a Möbius transformation, the condition of fixed points is

$$\frac{\alpha z + \beta}{\gamma z + \delta} = z \tag{3.60}$$

or equivalently

$$\gamma z^2 - (\alpha - \delta)z - \beta = 0 \tag{3.61}$$

If $\gamma \neq 0$, this is a quadratic equation and has at most two solutions. If $\gamma = 0$, this has one solution, unless $\alpha = \delta$. If $\gamma = 0$ and $\alpha = \delta$, there is no fixed point, unless $\beta = 0$. Finally, if $\gamma = \beta = 0$ and $\alpha = \delta \neq 0$, the mapping is the identity mapping. Thus, the following lemma has been proved.

Lemma 3.1 (Fixed points of a Möbius transformation) *A Möbius transformation, other than the identity mapping, has at most two fixed points. If a Möbius transformation has three or more fixed points, it must be the identity mapping.*

Example 3.15 Let us find the Möbius transformation with fixed points $z = \pm 2$. Substituting $z = 2$ into Eq. 3.60 yields

$$2\alpha + \beta = 2(2\gamma + \delta)$$

Similarly, for $z = -2$, it follows that

$$-2\alpha + \beta = -2(-2\gamma + \delta)$$

From the two equations, $\alpha = \delta$ and $\beta = 4\gamma$, and it follows that

$$w = \frac{\alpha z + 4\gamma}{\gamma z + \alpha} \tag{3.62}$$

which is the Möbius transformation with the fixed points at $z = \pm 2$.

Example 3.16 Let us find the Möbius transformation with fixed points $z = \pm 2$ and $z = 3$. From Example 3.15, the fixed points $z = \pm 2$ yields the transformation given by Eq. 3.62 and the fixed point $z = 3$ must satisfy

$$\frac{3\alpha + 4\gamma}{3\gamma + \alpha} = 3$$

which yields $\gamma = 0$. Then, the transformation becomes $w = z$, that is, the identity mapping.

3.4.3 Cross Ratio

Let us consider the following Möbius transformation:

$$F(z) = \frac{z - z_1}{z - z_3} \frac{z_2 - z_3}{z_2 - z_1} \tag{3.63}$$

It is clear that this transformation maps z_1, z_2, and z_3 onto 0, 1, and ∞, respectively. Suppose there is another Möbius transformation $G(z)$ which maps z_1, z_2, and z_3 onto 0, 1, and ∞, respectively. Then, the composition $G^{-1}(F(z))$ is a mapping with three fixed points, z_1, z_2, and z_3. According to Lemma 3.1, $G^{-1}(F(z))$ must be the identity mapping, which implies that the transformation G is identical to F. This illustrates the following lemma.

Lemma 3.2 (Uniqueness) *The unique Möbius transformation that respectively maps z_1, z_2, and z_3 onto 0, 1, and ∞ is given by Eq. 3.63.*

The image of z_4 under the mapping $F(z)$ is

$$F(z_4) = \frac{z_4 - z_1}{z_4 - z_3} \frac{z_2 - z_3}{z_2 - z_1} \tag{3.64}$$

which is called the cross ratio of the four complex numbers z_1, z_2, z_3, and z_4, denoted by (z_1, z_2, z_3, z_4). The cross ratio has a unique property useful for deriving desired transformations, as discussed below.

Suppose the points z_1, z_2, z_3, and z_4 are respectively mapped onto the points w_1, w_2, w_3, and w_4 under a Möbius transformation f. Then, $F(f^{-1}(w))$ maps w_1, w_2, and w_3 onto 0, 1, and ∞ and by definition $(w_1, w_2, w_3, w_4) = F(f^{-1}(w_4))$. Since $F(f^{-1}(w_4)) = F(z_4) = (z_1, z_2, z_3, z_4)$, it follows that

$$(w_1, w_2, w_3, w_4) = (z_1, z_2, z_3, z_4) \tag{3.65}$$

that is, the cross ratio of four points is invariant under Möbius transformations. These observations lead to the following theorem.

Theorem 3.3 (Implicit formula of a Möbius transformation) *There exists a unique Möbius transformation $w = f(z)$ that maps three distinct points z_1, z_2, and z_3 onto three distinct points w_1, w_2, and w_3, respectively. An implicit formula for the mapping is given by*

$$\frac{w - w_1}{w - w_3} \frac{w_2 - w_3}{w_2 - w_1} = \frac{z - z_1}{z - z_3} \frac{z_2 - z_3}{z_2 - z_1} \tag{3.66}$$

If one of those points is the point at infinity, ∞, the quotient of the two differences containing ∞ is replaced by 1.

Proof The right-hand side of Eq. 3.66 is the Möbius transformation $(z_1, z_2, z_3, z) = F(z)$, which maps z_1, z_2, and z_3 onto 0, 1, and ∞, respectively. The left-hand side of Eq. 3.66 is the Möbius transformation $(w_1, w_2, w_3, w) = G(w)$, which maps w_1, w_2, and w_3 onto 0, 1, and ∞, respectively.

It follows that the transformation $G^{-1}(F(z))$ maps z_1, z_2, and z_3 onto w_1, w_2, and w_3, and thus, $f(z) = G^{-1}(F(z))$. According to Lemma 3.2, F and G are both unique, and thus, $f(z) = G^{-1}(F(z))$ is also unique.   □

Example 3.17 Let us confirm that Eq. 3.66 maps z_1, z_2, and z_3 onto w_1, w_2, and w_3, respectively. Equation 3.66 is rewritten as

$$(z - z_3)(w - w_1)(z_2 - z_1)(w_2 - w_3) = (z - z_1)(w - w_3)(z_2 - z_3)(w_2 - w_1) \tag{3.67}$$

If $z = z_1$, the right-hand side of Eq. 3.67 is zero, and it follows that $w = w_1$. Similarly, if $z = z_3$, the left-hand side is zero, and thus, $w = w_3$. If $z = z_2$, Eq. 3.67 reduces to $(w - w_1)(w_2 - w_3) = (w - w_3)(w_2 - w_1)$ which has a solution $w = w_2$.

Although a Möbius transformation, Eq. 3.53, includes four constants, α, β, γ, and δ, its specific form can be determined by the ratios of three of these constants to the fourth. This implies that three conditions determine a unique transformation, which is consistent with Theorem 3.3. Equation 3.66 can be used to find specific Möbius transformations that map three points onto three other points.

Example 3.18 Let us find the Möbius transformation that maps $z_1 = 0$, $z_2 = 1$, and $z_3 = \infty$ onto $w_1 = -1$, $w_2 = -i$, and $w_3 = 1$, respectively. Substituting these values into Eq. 3.66, it follows that

$$\frac{w + 1}{w - 1} \frac{-i - 1}{-i + 1} = \frac{z - 0}{z - \infty} \frac{1 - \infty}{1 - 0} = z$$

where $(1 - \infty)/(z - \infty)$ is replaced by 1. Solving this equation for w yields

$$w = \frac{z - i}{z + i}$$

which is the Cayley transformation, discussed in Example 3.14.

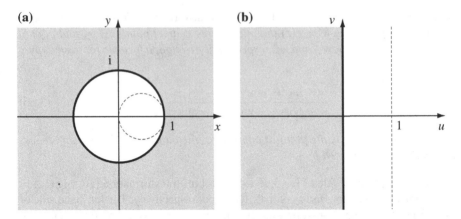

Fig. 3.33 Mapping by $w = -(z+1)/(z-1)$. **a** Unit disk and a circle passing through the origin in the z plane. **b** Right half plane and $w = 1 + iv$ in the w plane

Example 3.19 Let us find the Möbius transformation that maps the unit disk (the interior of the unit circle) in the z plane onto the right half of the w plane, as shown in Fig. 3.33.

Consider the mapping of $z_1 = -1$, $z_2 = i$, and $z_3 = 1$ on the unit circle (a thick line in Fig. 3.33a) onto $w_1 = 0$, $w_2 = i$, and $w_3 = \infty$ on $u = 0$ (a thick line in Fig. 3.33b), respectively. Substituting these values into Eq. 3.66, it follows that

$$\frac{w-0}{w-\infty}\frac{i-\infty}{i-0} = \frac{w}{i} = \frac{z+1}{z-1}\frac{i-1}{i+1}$$

where $(i-\infty)/(w-\infty)$ is replaced by 1. Solving this equation for w yields

$$w = -\frac{z+1}{z-1} \qquad (3.68)$$

which is the desired transformation.

Example 3.20 Let us confirm that Eq. 3.68 maps the unit disk $|z| < 1$ onto the right half of the w plane. Using Eq. 3.54, the inverse of Eq. 3.68 is

$$z = \frac{-w+1}{-w-1} = \frac{w-1}{w+1}$$

For the unit circle in the z plane, $|z| = 1$, the images of points in the w plane satisfy

$$|w-1| = |w+1|$$

Squaring both sides gives

$$(u-1)^2 + v^2 = (u+1)^2 + v^2$$

which yields $u = 0$, the equation of the v axis.

The unit circle, $|z| = 1$, divides the z plane into two portions, the interior and exterior of the circle, and its image is the v axis, which divides the w plane into two portions, the right and left halves. The image of the point $z = 0$ is $w = 1$, and thus, the interior of the unit circle in the z plane is mapped onto the right half of the w plane.

For instance, the vertical line in the w plane passing through $u = 1$ (a dashed line in Fig. 3.33b), that is, $w = 1 + iv$, is inversely mapped as

$$z = x + iy = \frac{iv}{2 + iv}$$

which gives

$$\begin{cases} 2x - vy = 0 \\ vx + 2y - v = 0 \end{cases}$$

Eliminating v yields

$$\left(x - \frac{1}{2}\right)^2 + y^2 = \frac{1}{4}$$

which is the circle of radius $1/2$ with its center at $z = 1/2$ (a dashed line in Fig. 3.33a).

Example 3.21 Let us find the Möbius transformation that maps the unit disk (the interior of the unit circle) in the z plane onto the unit disk (the interior of the unit circle) in the w plane while the point $z = -0.5$ is mapped onto the center $w = 0$, as shown in Fig. 3.34.

Consider the mapping of $z_1 = -1, z_2 = 1$, and $z_3 = -0.5$ onto $w_1 = -1, w_2 = 1$, and $w_3 = 0$, respectively. Substituting these values into Eq. 3.66, it follows that

$$\frac{w + 1}{w} \frac{1}{1 + 1} = \frac{z + 1}{z + 0.5} \frac{1 + 0.5}{1 + 1}$$

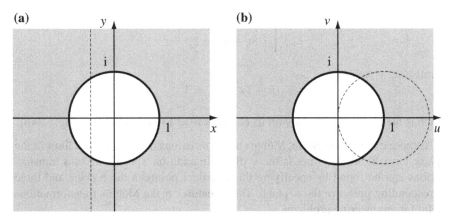

(a) **(b)**

Fig. 3.34 Mapping by $w = (2z + 1)/(z + 2)$. **a** Unit disk and $z = -0.5 + iy$ in the z plane. **b** Unit disk and a circle passing through the origin in the w plane

Solving this equation for w yields

$$w = \frac{2z + 1}{z + 2} \tag{3.69}$$

which is the desired transformation.

Example 3.22 Let us confirm that Eq. 3.69 maps the unit disk in the z plane onto the unit disk in the w plane while the point $z = -0.5$ is mapped onto the center $w = 0$. It is obvious that substituting $z = -0.5$ into Eq. 3.69 gives $w = 0$, which confirms the second condition. As for the first condition, for the unit circle in the w plane, $|w| = 1$, the corresponding points in the z plane satisfy

$$|2z + 1| = |z + 2|$$

Squaring both sides gives
$$x^2 + y^2 = 1$$

which is the equation of the unit circle, $|z| = 1$.

The unit circle, $|z| = 1$, divides the z plane into two portions, the interior and exterior of the circle, and its image is the unit circle, which divides the w plane into two portions, the interior and exterior of the circle. The image of the point $z = -0.5$ is $w = 0$, and thus, the interior of the unit circle in the z plane is mapped onto the interior of the unit circle in the w plane.

The vertical line in the z plane passing through $x = -0.5$ (a dashed line in Fig. 3.34a), that is, $z = -0.5 + iy$, is mapped as

$$w = u + iv = \frac{2yi}{1.5 + iy}$$

which gives

$$\begin{cases} 1.5u - vy = 0 \\ 1.5v + uy = 2y \end{cases}$$

Eliminating y yields

$$(u - 1)^2 + v^2 = 1$$

which is the circle of radius 1 with its center at $w = 1$ (a dashed line in Fig. 3.34b).

As these examples indicate, Möbius transformations map circles and lines in the z plane onto circles and lines in the w plane. In addition, specific Möbius transformations can be found by specifying three distinct points in the z plane and three corresponding points in the w plane. These features make Möbius transformations useful for engineering problems.

Example 3.23 Let us consider flow around a circular pond (of radius 1 with its center at the origin) near a shoreline (along $x = -2$), as shown in Fig. 3.35a. The velocity

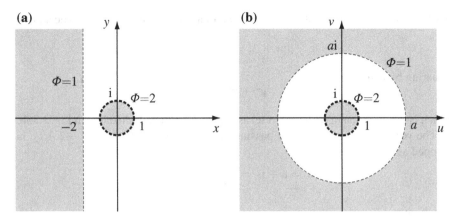

Fig. 3.35 Mapping by $w = (az + 1)/(z + a)$. **a** Circular pond of radius 1 with its center at the origin and shoreline along $x = -2$ in the z plane. **b** Concentric circles of radii 1 and a in the w plane

potential along the boundary of the pond (a thick dashed line in Fig. 3.35a) is maintained at $\Phi = 2$ and that along the shoreline (a dashed line in Fig. 3.35a) is constant at $\Phi = 1$.

To transform the flow domain ($|z| > 1$ and $x > -2$) in the z plane onto the annulus ($1 < |w| < a$) in the w plane, as shown in Fig. 3.35b, let us find the Möbius transformation that maps the unit disk onto the unit disk while the point $z = \infty$ is mapped onto $w = a$. Consider the mapping of $z_1 = -1$, $z_2 = 1$, and $z_3 = \infty$ onto $w_1 = -1$, $w_2 = 1$, and $w_3 = a$, respectively. Substituting these values into Eq. 3.66, it follows that

$$\frac{w+1}{w-a}\frac{1-a}{1+1} = \frac{z+1}{z-\infty}\frac{1-\infty}{1+1} = \frac{z+1}{2}$$

where $(1 - \infty)/(z - \infty)$ is replaced by 1. Solving this equation for w yields

$$w = \frac{az + 1}{z + a} \tag{3.70}$$

To determine a, consider the mapping of $z = -2$ onto $w = -a$, that is,

$$-a = \frac{-2a + 1}{-2 + a}$$

which has the solutions $a = 2 \pm \sqrt{3}$, of which $a = 2 + \sqrt{3}$ is valid since a is larger than 1 as seen in Fig. 3.35b.

Now, let us confirm that Eq. 3.70 with $a = 2 + \sqrt{3}$ is the desired transformation. Using Eq. 3.54, the inverse of Eq. 3.70 is

$$z = \frac{aw - 1}{-w + a}$$

For the unit circle in the z plane, $|z| = 1$, the images of points in the w plane satisfy

$$|aw - 1| = |-w + a|$$

Squaring both sides gives

$$u^2 + v^2 = 1$$

which is the equation of the unit circle, as shown by a thick dashed line in Fig. 3.35b.

The vertical line in the z plane passing through $x = -2$, that is, $z = -2 + iy$, is mapped as

$$w = u + iv = \frac{-2a + 1 + iay}{-2 + a + iy}$$

which gives

$$\begin{cases} \sqrt{3}u - vy = -3 - 2\sqrt{3} \\ \sqrt{3}v + uy = (2 + \sqrt{3})y \end{cases}$$

Eliminating y yields

$$u^2 + v^2 = (2 + \sqrt{3})^2$$

which is the circle of radius $a = 2 + \sqrt{3}$ with its center at the origin, as shown by a dashed line in Fig. 3.35b. Hence, Eq. 3.70 with $a = 2 + \sqrt{3}$ is indeed the desired transformation.

By noting that the equipotential lines are concentric circles, it is readily deduced that the complex potential in the w plane has the form of

$$\Omega(w) = c_1 \ln w + c_2 \tag{3.71}$$

The boundary conditions are

$$\begin{cases} \Phi(1) = \text{Re}[\Omega(1)] = c_1 \ln 1 + c_2 = 2 \\ \Phi(a) = \text{Re}[\Omega(a)] = c_1 \ln a + c_2 = 1 \end{cases}$$

which yields $c_1 = -1/\ln a$ and $c_2 = 2$.

The complex potential in the z plane can be obtained as the composite function $\Omega(w(z))$ of the two mappings, Eqs. 3.70 and 3.71, and it follows that

$$\Omega(z) = \Omega(w(z)) = -\frac{1}{\ln a} \ln \left(\frac{az + 1}{z + a} \right) + 2 \tag{3.72}$$

where $a = 2 + \sqrt{3}$. Figure 3.36 shows the corresponding equipotential lines and streamlines.

(a)   **(b)**

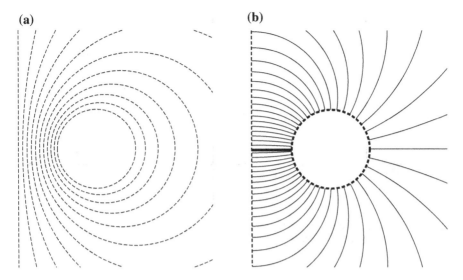

Fig. 3.36 Flow around a pond near a shoreline. **a** Equipotential lines. **b** Streamlines. (The thick line emanating from the pond to the shoreline corresponds to the branch cut and overlaps a streamline)

3.5 Schwarz–Christoffel Transformation

In some engineering problems, the domain of interest is often interior to a closed or open polygon. It becomes easier to obtain the solutions if the polygon and its interior can be mapped onto a simpler domain, such as the real axis and the upper half plane.

3.5.1 The Real Axis and a Polygon

Let us derive a mapping from the upper half of the z plane onto the interior of a polygon in the w plane. The rotational effect of a conformal mapping $w = f(z)$ at a point $z_0 = z(t_0)$ on a curve C is given by Eq. 3.5 as

$$\arg w'(t_0) = \arg f'(z_0) + \arg z'(t_0) \tag{3.73}$$

where parametric representations $z(t) = x(t) + iy(t)$ and $w(t) = f(z(t))$ are used. If C in the z plane is a segment of the x axis with positive sense to the right, then $\arg z'(t_0) = 0$ at each point $z_0 = z(t_0) = x_0$, and

$$\arg w'(t_0) = \arg f'(x_0) \tag{3.74}$$

This implies that if $f'(z)$ has a constant argument along the segment, $\arg w'(t)$ is constant. Therefore, in the w plane, the image C' of C is also a segment of a straight line.

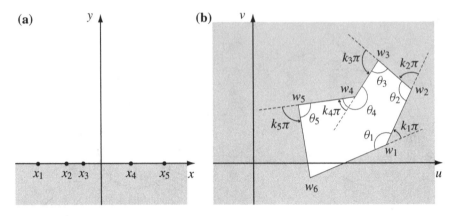

Fig. 3.37 Mapping by the Schwarz–Christoffel transformation. **a** Upper half of the z plane. **b** Interior of a closed polygon with vertices $w_1, w_2, \ldots, w_n$

Let us consider a polygon in the w plane, as shown in Fig 3.37, having n vertices at $w_j = f(x_j)$ that correspond to the points x_j on the x axis in the z plane, where

$$x_1 < x_2 < \cdots < x_{n-1} < x_n = \infty \tag{3.75}$$

and define the function f such that

$$f'(z) = \alpha(z - x_1)^{-k_1}(z - x_2)^{-k_2} \cdots (z - x_{n-1})^{-k_{n-1}} \tag{3.76}$$

where α is the complex constant and k_j are the real constants.

The argument of $f'(z)$ can be written as

$$\arg f'(z) = \arg \alpha - k_1 \arg(z - x_1)$$
$$- k_2 \arg(z - x_2) - \cdots - k_{n-1} \arg(z - x_{n-1}) \tag{3.77}$$

For $z = x < x_1$, $\arg(z - x_1) = \arg(z - x_2) = \cdots = \arg(z - x_{n-1}) = \pi$, and thus

$$\arg f'(z) = \arg \alpha - (k_1 + k_2 + \cdots + k_{n-1})\pi \tag{3.78}$$

For $x_1 < z = x < x_2$, $\arg(z - x_1) = 0$ and $\arg(z - x_2) = \cdots = \arg(z - x_{n-1}) = \pi$, and it follows that

$$\arg f'(z) = \arg \alpha - (k_2 + \cdots + k_{n-1})\pi \tag{3.79}$$

Therefore, as z moves to the right through the point $z = x_1$, $\arg f'(z)$ increases abruptly by the amount $k_1\pi$. Similarly, as z moves through the point $z = x_j$, the argument increases by $k_j\pi$.

In general, for $z = x$ between the points $z = x_{j-1}$ and x_j

$$\arg f'(z) = \arg \alpha - \sum_{v=j}^{n-1} k_v \pi \qquad (3.80)$$

which is constant, and it follows from Eq. 3.74 that w moves in the fixed direction w'_{j-1} along a straight line as z moves from x_{j-1} to x_j. The direction w'_{j-1} changes by the amount $k_j \pi$ as the point w moves along the segment $w_{j-1}w_j$ through the vertex w_j to the segment $w_j w_{j+1}$

$$\arg w'_j - \arg w'_{j-1} = k_j \pi \qquad (3.81)$$

where w_j is the image point of z_j.

The angle $k_j \pi$ in Eq. 3.81 is exterior to the polygon at the vertex w_j and satisfies the inequality $-\pi < k_j \pi < \pi$, and thus, $-1 < k_j < 1$. For a closed polygon, the sum of the exterior angles is equal to 2π, and the exterior angle at w_n, the image of $z_n = \infty$, is given by

$$k_n \pi = 2\pi - \sum_{v=1}^{n-1} k_v \pi \qquad (3.82)$$

and only $n - 1$ angles need to be specified.

If $\sum_{v=1}^{n-1} k_v \pi \leq \pi$, then $k_n \pi \geq \pi$ and the vertices w_j cannot form a closed polygon. In such a case, the vertex w_n is at infinity and a polygon in the w plane is open, as shown in Fig. 3.38. Infinite open polygons can be considered as limiting cases of closed polygons.

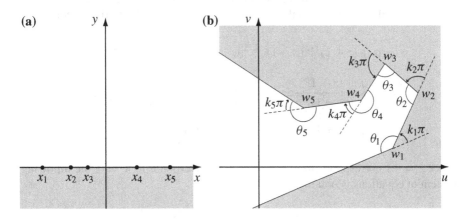

Fig. 3.38 Mapping by the Schwarz–Christoffel transformation. **a** Upper half of the z plane. **b** Interior of an open polygon with vertices $w_1, w_2, \ldots, \infty$

3.5.2 Transformation in Integral Form

It is geometrically obvious that $k_j \pi + \theta_j = \pi$, as seen in Figs. 3.37 and 3.38, and thus, k_j is given in terms of the interior angle θ_j as

$$k_j = 1 - \frac{\theta_j}{\pi} \tag{3.83}$$

Hence, Eq. 3.76 is rewritten as

$$\frac{dw}{dz} = \alpha(z - x_1)^{\theta_1/\pi - 1}(z - x_2)^{\theta_2/\pi - 1} \cdots (z - x_{n-1})^{\theta_{n-1}/\pi - 1} \tag{3.84}$$

and w can be expressed as an indefinite integral[2]

$$w = \alpha \int (z - x_1)^{\theta_1/\pi - 1}(z - x_2)^{\theta_2/\pi - 1} \cdots (z - x_{n-1})^{\theta_{n-1}/\pi - 1} dz \tag{3.85}$$

Equation 3.85 is known as the Schwarz–Christoffel transformation that maps the upper half of the z plane onto the interior of the polygon in the w plane and the real axis onto the boundary of the polygon. It must be noted, however, that the integral in Eq. 3.85 cannot always be evaluated in terms of elementary functions. Also, the integral often involves multi-valued functions, and a specific branch must be selected to fit adequately the conditions specified in the problem.

Example 3.24 Let us determine a function that maps the upper half plane onto the domain with a step along the segment $0 \le v \le 1$ in the w plane as shown in Fig. 3.39.

Let $x_1 = -1$ and $x_2 = 1$ be mapped onto $w_1 = i$ and $w_2 = 0$, respectively; then $\theta_1 = 3\pi/2$ and $\theta_2 = \pi/2$, and the Schwarz–Christoffel transformation becomes

$$w = \alpha \int (z + 1)^{1/2}(z - 1)^{-1/2} dz$$

$$= \alpha \int \frac{z + 1}{(z^2 - 1)^{1/2}} dz = \alpha \int \left[\frac{z}{(z^2 - 1)^{1/2}} + \frac{1}{(z^2 - 1)^{1/2}} \right] dz$$

$$= \alpha \left[(z^2 - 1)^{1/2} + \ln\left(z + (z^2 - 1)^{1/2}\right) \right] + \beta$$

Since $z = -1$ corresponds to $w = i$ and $z = 1$ corresponds to $w = 0$, the following system of equations is obtained:

$$\begin{cases} i = \alpha \ln(-1) + \beta = \pi \alpha i + \beta \\ 0 = \alpha \ln(1) + \beta = \beta \end{cases}$$

---

[2] Complex integration is discussed in Chap. 4. In this section, a simple extension of indefinite integrals from real-variable to complex-variable calculus is tentatively used.

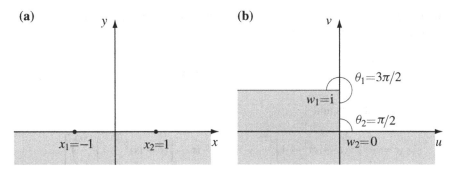

Fig. 3.39 Mapping by the Schwarz–Christoffel transformation. **a** Upper half of the z plane. **b** Domain above the boundary with a step along the segment $0 \leq v \leq 1$

which yields $\alpha = 1/\pi$ and $\beta = 0$. Thus, w becomes

$$w = \frac{1}{\pi} \left[(z^2 - 1)^{1/2} + \ln \left(z + (z^2 - 1)^{1/2} \right) \right]$$

which is the desired transformation.

3.5.3 Parametric Equations for Flow Profiles

Let us consider flow profiles in the w plane that are bounded by line segments. As discussed in Example 3.8, the equipotential lines and streamlines are the level curves, defined by

$$\begin{cases} \Phi = p \\ \Psi = s \end{cases} \tag{3.86}$$

for various values of p and s, and these lines are, respectively, parallel to the Ψ and Φ axes in the Ω plane, the profile of which is equivalent to uniform flow in the x direction in the z plane. By analogy, the Schwarz–Christoffel transformation of uniform flow in the z plane yields flow bounded by line segments in the w plane.

The images of equipotential lines $\Phi = p$ and streamlines $\Psi = s$ in the w plane can be found by the parametric representations

$$w = u + iv = u(p, t) + iv(p, t) \tag{3.87}$$

for equipotential lines and

$$w = u + iv = u(t, s) + iv(t, s) \tag{3.88}$$

for streamlines, where $-\infty < t < \infty$ is the parameter. In particular, the streamline with $s = 0$ corresponds to the boundary defined by the line segments.

Example 3.25 The flow profile around the step considered in Example 3.24 can be obtained by the parametric equations

$$w = \frac{1}{\pi}\left[\left\{(p+it)^2 - 1\right\}^{1/2} + \ln\left(p + it + \left\{(p+it)^2 - 1\right\}^{1/2}\right)\right]$$

for equipotential lines and

$$w = \frac{1}{\pi}\left[\left\{(t+is)^2 - 1\right\}^{1/2} + \ln\left(t + is + \left\{(t+is)^2 - 1\right\}^{1/2}\right)\right]$$

for streamlines. For various values of p and s, the corresponding points $w = (u, v)$ for $-\infty < t < \infty$ are located. Figure 3.40 shows the resultant flow profiles.

Example 3.26 Let us determine a function that maps the upper half plane onto the domain lying outside an isosceles triangle (wedge) with vertices $\pm b$ and ai in the w plane as shown in Fig. 3.41. The base angle is $k_1\pi$.

Let $x_1 = -1$, $x_2 = 0$, and $x_3 = 1$ be mapped onto $w_1 = -b$, $w_2 = ai$, and $w_3 = b$, respectively, then $\theta_1 = \theta_3 = \pi - k_1\pi$ and $\theta_2 = \pi + 2k_1\pi$, and the Schwarz–Christoffel transformation becomes

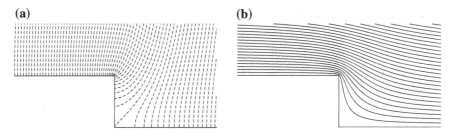

(a)   **(b)**

Fig. 3.40 Flow around a step. **a** Equipotential lines. **b** Streamlines

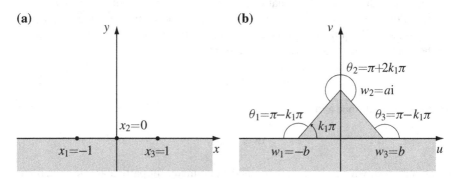

(a)   **(b)**

Fig. 3.41 Mapping by the Schwarz–Christoffel transformation. **a** Upper half of the z plane. **b** Domain outside the wedge with vertices $\pm b$ and ai

$$w = \alpha \int (z+1)^{-k_1} z^{2k_1} (z-1)^{-k_1} dz$$

$$= \alpha \int \frac{z^{2k_1}}{(z^2-1)^{k_1}} dz$$

to which a closed-form solution cannot always be found.

As a limiting case, let us consider $b \rightarrow 0$ and $k_1 \rightarrow 1/2$, which squashes the isosceles triangle down to the line segment $[0, ai]$. The resultant domain is the upper half of the w plane slit from 0 to ai as shown in Fig. 3.42.

The limiting equation becomes

$$w = \alpha \int \frac{z}{(z^2-1)^{1/2}} dz = \alpha(z^2-1)^{1/2} + \beta$$

Since $z = \pm 1$ corresponds to $w = 0$ and $z = 0$ corresponds to $w = ai$, the following system of equations is obtained:

$$\begin{cases} 0 = \alpha(1-1)^{1/2} + \beta = \beta \\ ai = \alpha(0-1)^{1/2} + \beta = \alpha i + \beta \end{cases}$$

which yields $\alpha = a$ and $\beta = 0$. Thus, w becomes

$$w = a(z^2-1)^{1/2}$$

which is the desired transformation.

Example 3.27 The flow profile around the vertical segment considered in Example 3.26 can be obtained by the parametric equations

$$w = a\left\{(p+it)^2 - 1\right\}^{1/2}$$

Fig. 3.42 Upper half of the w plane slit from 0 to ai

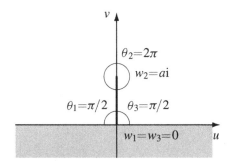

(a)   **(b)**

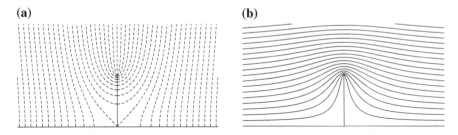

Fig. 3.43 Flow around a vertical segment. **a** Equipotential lines. **b** Streamlines

for equipotential lines and

$$w = a \left\{ (t + is)^2 - 1 \right\}^{1/2}$$

for streamlines. For various values of p and s, the corresponding points $w = (u, v)$ for $-\infty < t < \infty$ are located. Figure 3.43 shows the resultant flow profiles.

Chapter 4
Boundary Value Problems and Integration

Motivating Problem 5: Cut-Off Walls for Irrigation

Groundwater flows around right-angle impermeable faults with a flow velocity $q_u/A = 2$, as shown in Fig. 4.1a. There is a dry area in the central west region. For the purpose of irrigation in the dry area, an installation of cut-off walls is planned so that more groundwater flows into the area. Figure 4.1b shows the resultant configuration. Guiding cut-off walls (not shown) squeeze flow inlet and outlet into half the original widths and a cut-off wall between $(1, 0.5)$ and $(0.5, 1)$ forces flow toward the dry area.

Task 5-1   Draw the groundwater flow profile and evaluate the discharge in the dry area, represented by the amount flowing between $(0, 0.5)$ and $(0.5, 0.5)$, without a cut-off wall.

Task 5-2   Draw the groundwater flow profile and evaluate the discharge in the dry area with the cut-off walls.

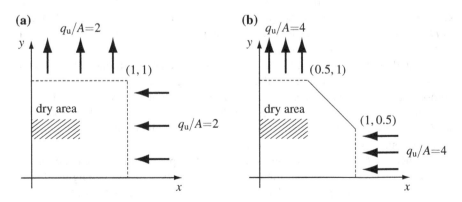

Fig. 4.1 Groundwater flowing through a dry area near right-angle impermeable faults. **a** Flow without a cut-off wall. **b** Flow with cut-off walls

© Springer International Publishing Switzerland 2015
K. Sato, *Complex Analysis for Practical Engineering*,
DOI 10.1007/978-3-319-13063-7_4

• **Solution Strategy to Motivating Problem 5**

Different physical situations (cut-off walls in this problem) impose different conditions at the boundaries of the problem domain. It is obvious that flow is affected by boundary conditions, and thus, a mathematical procedure needs to be established to determine solutions that satisfy the specific boundary conditions.

A possible procedure is conformal mapping, as discussed in Chap. 3. Indeed, Task 5-1 is readily solved by mapping with a power function. When the specifications of boundary conditions are not simple, however, finding appropriate mappings is not an easy task. To achieve versatility in a solution method, a numerical technique is developed in this chapter.

4.1 Boundary Value Problems

Engineering problems, when formulated mathematically, lead to partial differential equations, such as Laplace's equation and associated boundary conditions. The problem of finding a solution to the partial differential equation, which also satisfies the boundary conditions, is called the boundary value problem. There exist various kinds of boundaries, including infinite (as seen in previous chapters), finite, simply connected, and multiply connected boundaries.

4.1.1 Domain and Boundary

A δ neighborhood of a point z_0, as introduced in Sect. 2.3.1, is the set of all points z such that

$$|z - z_0| < \delta \tag{4.1}$$

where δ is any given positive number. A point z_0 is said to be an interior point of a set Π whenever there is some neighborhood z_0 that contains only points belonging to Π, as shown in Fig. 4.2. If there exists a neighborhood of z_0 containing no point of Π, z_0 is called an exterior point of Π.

A boundary point of a set Π is a point all of whose neighborhood contains both points that belong to Π and points that do not belong to Π. The set of all boundary points of a set Π is called the boundary of Π. The point b shown in Fig. 4.2 is a boundary point of Π and the totality of all boundary points Γ is the boundary. If a set Π contains none of its boundary points b, the set is called open. If a set contains all of its boundary points, the set is called closed.

An open set Π is said to be connected if any two points of the set can be joined by a path consisting of straight line segments all points of which belong to Π. An open set that is connected is called a domain. Figure 4.2 shows a domain Π, in which the points z_1 and z_2 are joined. A domain together with some or all of its boundary points (b or Γ) is called a region.

Fig. 4.2 Domain Π, containing joined points z_1 and z_2, and boundary Γ, formed by boundary points b

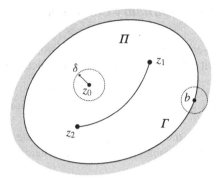

A closed path is a path such that its terminal point coincides with its initial point, and a simple closed path is a closed path that does not intersect or touch itself. If any simple closed path in a domain Π can be shrunk to a point without leaving Π, the domain Π is called a simply connected domain. As shown in Fig. 4.3a, a simple closed path C can be shrunk to a point that lies in Π, and thus the domain is simply connected. On the other hand, for the domain shown in Fig. 4.3b, there is a simple closed path C lying in Π which cannot be shrunk to a point without leaving Π, and the domain is said to be multiply connected.

By convention, the positive direction on the boundary is selected so that when facing in the positive direction the domain of interest lies to the left. Hence, the counterclockwise direction is the positive direction for the outer boundary Γ, as shown in Fig. 4.3. For the inner boundaries Γ_1 and Γ_2 for the multiply connected domain, in contrast, the clockwise direction is the positive direction.

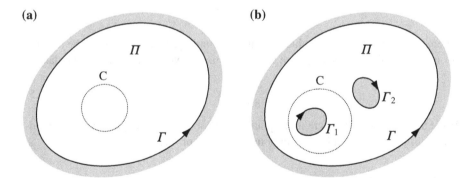

Fig. 4.3 Domains of different types. **a** Simply connected domain, with a closed path C that can be shrunk to an interior point. **b** Multiply connected domain, with a closed path C that cannot be shrunk to an interior point

4.1.2 Boundary Conditions in Engineering Problems

In engineering problems, the two most common kinds of boundary conditions are Dirichlet and Neumann types, which are briefly reviewed.

4.1.2.1 Dirichlet Boundary Condition

Along the Dirichlet-type boundary, the value of the function of interest is prescribed. In flow problems, the velocity potential Φ is prescribed as

$$\Phi = P(\zeta) \tag{4.2}$$

where $P(\zeta)$ is the boundary value at ζ on the boundary. A special case is that the boundary is an equipotential line:

$$\Phi = \text{constant} \tag{4.3}$$

In the theory of partial differential equations, a problem with only this type of boundary condition is called a Dirichlet boundary value problem.

4.1.2.2 Neumann Boundary Condition

Along the Neumann-type boundary, the derivative of the function of interest normal to the boundary is prescribed. In flow problems, the flux normal to the boundary surface, which is proportional to $\partial\Phi/\partial n$, is prescribed

$$\frac{\partial\Phi}{\partial n} = S(\zeta) \tag{4.4}$$

where $S(\zeta)$ is the boundary value at ζ on the boundary. A special case is the no-flow (or impermeable) boundary where the flux vanishes everywhere on the boundary:

$$\frac{\partial\Phi}{\partial n} = 0 \tag{4.5}$$

In the theory of partial differential equations, a problem with only this type of boundary condition is called a Neumann boundary value problem.

4.1.2.3 Some Remarks

If the Dirichlet condition applies over a part of the boundary Γ_P and the Neumann condition applies over the remaining part Γ_S, as shown in Fig. 4.4, where $\Gamma_P \cup \Gamma_S = \Gamma$ and $\Gamma_P \cap \Gamma_S = \emptyset$, such a problem is called a mixed boundary value problem.

Fig. 4.4 Boundary
conditions prescribing Φ on
the Dirichlet-type boundary
Γ_P and $\partial\Phi/\partial n$ on the
Neumann-type boundary Γ_S

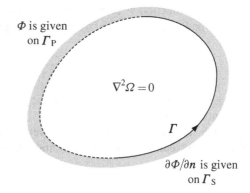

Φ is given
on Γ_P

$\nabla^2\Omega = 0$

Γ

$\partial\Phi/\partial n$ is given
on Γ_S

Dirichlet and mixed boundary value problems in a finite domain have unique
solutions, while Neumann boundary value problems have solutions within an arbitrary additive constant. To ensure the uniqueness of the solution for the Neumann
problem, a reference value of Φ must be set at a point on the boundary.

In a well-posed boundary value problem, along each part of the boundary one and
only one condition is prescribed in order to define a solution fully; hence, boundary
values $\partial\Phi/\partial n$ and Φ are left unknown on the Dirichlet-type boundary Γ_P and the
Neumann-type boundary Γ_S, respectively.

4.1.3 Boundary Conditions in Complex Analysis

Since the velocity potential Φ is the real part of the complex potential Ω, the Dirichlet
boundary condition given by Eq. 4.2 can readily be handled in complex analysis. In
contrast, $\partial\Phi/\partial n$ is not included in Ω, and the Neumann boundary condition given
by Eq. 4.4 must be treated in a special manner.

As shown in Fig. 4.5, the outward normal and tangent vectors, **n** and **u**, along the
boundary are perpendicular to each other and form a Cartesian coordinate system.

Fig. 4.5 Normal and tangent
vectors **n** and **u** forming a
Cartesian coordinate system

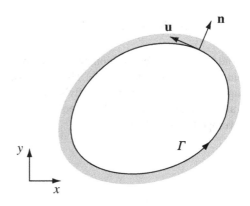

The stream function Ψ is the harmonic conjugate of the velocity potential Φ, and thus, Φ and Ψ can be related by the following Cauchy–Riemann equation:

$$\frac{\partial \Phi}{\partial n} = \frac{\partial \Psi}{\partial u} \tag{4.6}$$

which implies that the Neumann boundary condition $\partial \Phi / \partial n$ can be replaced by the tangential gradient, $\partial \Psi / \partial u$.

For instance, if $\partial \Phi / \partial n = \sigma$ is given on the boundary Γ_S, then according to Eq. 4.6, the Neumann boundary condition can be expressed in terms of Ψ as

$$\Psi = \sigma s + \Psi_0 \tag{4.7}$$

where s is the length from an arbitrary reference point measured along the boundary and Ψ_0 is the reference stream function at the reference point. In particular, as a commonly encountered Neumann boundary condition, an impermeable boundary is given by $\partial \Phi / \partial n = 0$ (Eq. 4.5), and the corresponding condition in terms of Ψ becomes

$$\Psi = \text{constant} \tag{4.8}$$

In this way, the Neumann boundary condition can be handled with the stream function in complex analysis.

In a well-posed boundary value problem, only half of the boundary variables is prescribed. The boundary conditions are specified in terms of Φ and Ψ as

$$\begin{cases} \Phi(\zeta) = P(\zeta) & (\zeta \in \Gamma_P) \\ \Psi(\zeta) = S(\zeta) & (\zeta \in \Gamma_S) \end{cases} \tag{4.9}$$

where $\Gamma_P \cup \Gamma_S = \Gamma$ and $\Gamma_P \cap \Gamma_S = \emptyset$. The boundary values Ψ on the Dirichlet-type boundary Γ_P and Φ on the Neumann-type boundary Γ_S are unknown and must be determined for the solution of a boundary value problem.

4.2 Complex Integration

In addition to differentiation discussed in Sect. 2.3, integration is another important topic in complex calculus. The theory of integration is of great use in practical engineering. In particular, Cauchy's integral theorem and its implied Cauchy's integral formula provide sound mathematical foundations for a numerical scheme to be developed in this chapter.

4.2.1 Contours

Complex integration is defined on paths in the complex plane. To represent a path in the plane, the parametric equation

$$z(t) = x(t) + iy(t) \tag{4.10}$$

where t is the parameter, is of use. If $x(t)$ and $y(t)$ are continuous, a set of points $z(t)$ is called an arc. The points $z = z(t)$ are ordered according to increasing values of the parameter t, and the arc is given an orientation.

As discussed in Sect. 4.1.1, an arc C is simple if it does not intersect or touch itself; $z(t_1) \neq z(t_2)$ whenever $t_1 \neq t_2$. For $a \leq t \leq b$, an arc C with $z(b) = z(a)$ is closed. If $z(b) = z(a)$ is the only point of intersection, C is a simple closed curve. As the parameter t increases from a to b, the point $z(t)$ moves from the initial point $z(a)$ to the terminal point $z(b)$ along the arc C. The opposite arc $-C$ consists of the same set of points but with the reversed order, which is given by

$$z(t) = x(-t) + iy(-t) \tag{4.11}$$

for $-b \leq t \leq -a$. The arc $-C$ is traversed from $z(b)$ to $z(a)$.

Example 4.1 The unit circle (the circle of radius 1 and its center at the origin) given by

$$z(\theta) = e^{i\theta}$$

for $0 \leq \theta \leq 2\pi$ forms a simple closed curve about the origin. The orientation is in the counterclockwise direction.

Example 4.2 The unit circle given by

$$z(\theta) = e^{-i\theta}$$

for $-2\pi \leq \theta \leq 0$ forms a simple closed curve about the origin, as is the case with the example above. The orientation, however, is reversed; the circle is traversed in the clockwise direction.

When the derivatives of $x(t)$ and $y(t)$ in Eq. 4.10 exist and are continuous throughout the entire interval $a \leq t \leq b$, the arc C is called a differentiable arc. The derivative

$$z'(t) = x'(t) + iy'(t) \tag{4.12}$$

gives a tangent vector at each point $z(t)$ on the arc C. If z' is continuous and nonzero on the interval, the arc C is said to be smooth.

A contour is an arc constructed by joining a finite number of smooth arcs end to end. Let C_1, C_2, $\ldots$, C_n be n smooth arcs such that the initial point of C_j

coincides with the terminal point of C_{j-1}, for $j = 2, 3, \ldots, n$. Then, the contour C
is expressed as

$$C = C_1 + C_2 + \cdots + C_n \tag{4.13}$$

and is piecewise smooth.

If the terminal point coincides with the initial point and that point is the only
point of intersection, a contour C is called a simple closed contour. The points on
any simple closed contour C are boundary points defined in Sect. 4.1.1.

4.2.2 Contour Integrals

Let $f(z)$ be a continuous function at all points of a contour C in the complex plane.
Subdividing C by points $z_0, z_1, z_2, \ldots, z_n$ yields n arcs as shown in Fig. 4.6. The
contour C is oriented in the direction from z_0 to z_n. On each arc joining z_{j-1} to z_j,
let us choose an arbitrary point ζ_j. The sum of $f(\zeta_j)(z_j - z_{j-1})$ for $j = 1, 2, \ldots, n$
can be formed as

$$S_n = f(\zeta_1)(z_1 - z_0) + f(\zeta_2)(z_2 - z_1) + \cdots + f(\zeta_n)(z_n - z_{n-1})$$

$$= \sum_{j=1}^{n} f(\zeta_j) \Delta z_j \tag{4.14}$$

where $\Delta z_j = z_j - z_{j-1}$.

If the number of subdivision n increases in such a way that the greatest $|\Delta z_j|$
approaches zero, the sum S_n approaches a limiting value which is independent of the
mode of subdivision. The limiting value of S_n as $n \to \infty$ is called the complex line
integral or contour integral of $f(z)$ over the path of integration C, denoted by

$$\int_C f(z) dz = \lim_{n \to \infty} \sum_{j=1}^{n} f(\zeta_j) \Delta z_j \tag{4.15}$$

Fig. 4.6 Subdivision points
z_j and evaluation points ζ_j
along the contour C from z_0
to z_n

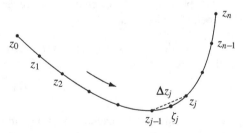

When a contour is closed, it is often written as

$$\oint_C f(z)dz = \lim_{n \to \infty} \sum_{j=1}^{n} f(\zeta_j)\Delta z_j \tag{4.16}$$

where the contour is oriented in the positive sense such that the interior of C is kept on the left.

From the definition mentioned above, the following properties of integrals are directly implied.

Linearity If the integrals of f and g over a path C exist, the integral of $c_1 f + c_2 g$, where c_1 and c_2 are constants, also exists and

$$\int_C (c_1 f + c_2 g)dz = c_1 \int_C f dz + c_2 \int_C g dz \tag{4.17}$$

Reversing limits of integration If the integral of f over a path C from z_0 to z_n exists, reversing the contour from z_n to z_0 (with the same path) introduces a minus sign as

$$\int_C f dz = \int_{z_0}^{z_n} f dz = -\int_{-C} f dz = -\int_{z_n}^{z_0} f dz \tag{4.18}$$

Partitioning path of integration If the integral of f over a path C exists and $C = C_1 + C_2$, then

$$\int_C f dz = \int_{C_1} f dz + \int_{C_2} f dz \tag{4.19}$$

Finally, an important inequality relating to complex integration is introduced. If $f(z)$ is continuous on the contour C and M is an upper bound for the modulus of $f(z)$ on C, that is, $|f(z)| \leq M$, then

$$|S_n| = \left| \sum_{j=1}^{n} f(\zeta_j)\Delta z_j \right| \leq \sum_{j=1}^{n} |f(\zeta_j)||\Delta z_j| \leq M \sum_{j=1}^{n} |\Delta z_j| \tag{4.20}$$

where the sum of $|\Delta z_j|$ approaches the length L of the contour C as $n \to \infty$. It follows that the modulus of the integral of $f(z)$ along C cannot exceed ML, that is,

$$\left| \int_C f(z)dz \right| \leq ML \tag{4.21}$$

which is known as the *ML* inequality.

4.2.3 Definite Integrals

In practice, Eq. 4.15 is too cumbersome to apply to evaluate definite integrals. Two kinds of evaluation methods are considered.

4.2.3.1 Parameterization

Let $f(z)$ be a continuous function on a piecewise smooth contour C. If C is represented by $z = z(t)$, where $a \leq t \leq b$, the definite integral is obtained as

$$\int_C f(z)dz = \int_a^b f(z(t))z'(t)dt \tag{4.22}$$

Note that $f(z)$ needs to be continuous on C but not necessarily analytic. In general, the definite integral depends not only on the endpoints (a and b) of the contour C but also on the contour itself.

Example 4.3 Let us evaluate the definite integral of $f(z) = \operatorname{Re} z = x$ from 0 to $1 + i$ along the contour OAB and along OB, as shown in Fig. 4.7.

The arc OA is given by $z(t) = t$ for $0 \leq t \leq 1$, and it follows that

$$\begin{cases} f(z(t)) = \operatorname{Re} z = t \\ z'(t) = 1 \end{cases}$$

Similarly, the arc AB is given by $z(t) = 1 + it$ for $0 \leq t \leq 1$, and it follows that

$$\begin{cases} f(z(t)) = \operatorname{Re} z = 1 \\ z'(t) = i \end{cases}$$

Fig. 4.7 Contours OAB and OB from 0 to $1 + i$

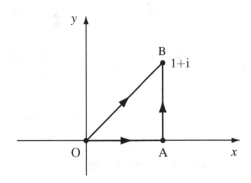

According to Eq. 4.19, the definite integral along the contour OAB is obtained as

$$\int_{OAB} f(z)dz = \int_{OA} f(z)dz + \int_{AB} f(z)dz$$

$$= \int_0^1 t\,dt + \int_0^1 i\,dt = \frac{1}{2}t^2\Big|_0^1 + it\Big|_0^1$$

$$= \frac{1}{2} + i$$

On the other hand, the contour OB is given by $z(t) = t + it$ for $0 \le t \le 1$, and it follows that

$$\begin{cases} f(z(t)) = \text{Re}\, z = t \\ z'(t) = 1 + i \end{cases}$$

The definite integral along the contour OB is obtained as

$$\int_{OB} f(z)dz = \int_0^1 t(1 + i)dt = \frac{1+i}{2}t^2\Big|_0^1$$

$$= \frac{1}{2} + \frac{1}{2}i$$

which differs from the result along the contour OAB.

Example 4.4 Let us evaluate the definite integral of $f(z) = z$ from 0 to $1 + i$ along the contour OAB and along OB, as shown in Fig. 4.7. The arc OA is given by $z(t) = t$ for $0 \le t \le 1$, and it follows that

$$\begin{cases} f(z(t)) = z = t \\ z'(t) = 1 \end{cases}$$

Similarly, the arc AB is given by $z(t) = 1 + it$ for $0 \le t \le 1$, and it follows that

$$\begin{cases} f(z(t)) = z = 1 + it \\ z'(t) = i \end{cases}$$

According to Eq. 4.19, the definite integral along the contour OAB is obtained as

$$\int_{OAB} f(z)dz = \int_{OA} f(z)dz + \int_{AB} f(z)dz$$

$$= \int_0^1 tdt + \int_0^1 (1+it)idt = \frac{1}{2}t^2\Big|_0^1 + \left(t+i\frac{t^2}{2}\right)i\Big|_0^1$$

$$= i$$

The contour OB is given by $z(t) = t + it$ for $0 \le t \le 1$, and it follows that

$$\begin{cases} f(z(t)) = z = t + it \\ z'(t) = 1 + i \end{cases}$$

The definite integral along the contour OB is obtained as

$$\int_{OB} f(z)dz = \int_0^1 (t+it)(1+i)dt = 2i\int_0^1 tdt = it^2\Big|_0^1$$

$$= i$$

which is the same as the result along the contour OAB.

Example 4.5 Let us evaluate the definite integral of $f(z) = 1/z$ over the unit circle C in the counterclockwise direction. The contour C is given by $z(\theta) = e^{i\theta}$ for $0 \le \theta \le 2\pi$, and it follows that

$$\begin{cases} f(z(\theta)) = e^{-i\theta} \\ z'(\theta) = ie^{i\theta} \end{cases}$$

The definite integral along the contour C is obtained as

$$\int_C f(z)dz = \int_0^{2\pi} e^{-i\theta}ie^{i\theta}d\theta = i\int_0^{2\pi} d\theta$$

$$= 2\pi i$$

If the orientation is reversed, the contour is given by $z(\theta) = e^{-i\theta}$ for $-2\pi \le \theta \le 0$, and it follows that

$$\begin{cases} f(z(\theta)) = e^{i\theta} \\ z'(\theta) = -ie^{-i\theta} \end{cases}$$

The definite integral along the reversed contour is obtained as

$$\int_{-C} f(z)dz = \int_{-2\pi}^{0} e^{i\theta}(-ie^{-i\theta})d\theta = -i\int_{-2\pi}^{0} d\theta$$

$$= -2\pi i$$

which is the negative of the result for the original contour.

4.2.3.2 Indefinite Integration

Let $f(z)$ be analytic in a simply connected domain. There exists an indefinite integral $F(z)$ of $f(z)$ in the same domain, such that $F'(z) = f(z)$; of necessity, $F(z)$ is also analytic. For all paths C in the domain joining two points z_0 and z_n, the definite integral is obtained as follows:

$$\int_C f(z)dz = \int_{z_0}^{z_n} f(z)dz = F(z_n) - F(z_0) \tag{4.23}$$

This is the analog of the evaluation of definite integrals in real-variable calculus. It should be noted that, if the integrand is analytic in a simply connected domain, the value of definite integral from z_0 to z_n is independent of path, as proved in the next section.

Example 4.6 Let us revisit Examples 4.3 and 4.4. Since $f(z) = \text{Re}\, z = x$ is not analytic, Eq. 4.23 cannot be applied to Example 4.3. On the other hand, $f(z) = z$ and its indefinite integral $z^2/2$ are analytic in the entire complex plane, which is certainly simply connected, and thus, Eq. 4.23 can be applied to Example 4.4:

$$\int_0^{1+i} zdz = \frac{1}{2}z^2\Big|_0^{1+i} = \frac{(1+i)^2}{2} = i$$

Example 4.7 Let us evaluate the definite integral of $1/z$ from $-i$ to i. If the complex plane without the origin and the negative real axis is taken for a simply connected domain, the integrand $1/z$ and its indefinite integral $\text{Ln}\,z$ are analytic in such a domain, and thus, Eq. 4.23 can be applied to evaluate the definite integral:

$$\int_{-i}^{i} \frac{1}{z}dz = \text{Ln}\,z\Big|_{-i}^{i} = \text{Ln}(i) - \text{Ln}(-i) = \frac{\pi}{2}i - \left(-\frac{\pi}{2}i\right) = \pi i$$

Example 4.8 Let us evaluate the definite integral of $1/z$ over the unit circle. Any simply connected domain containing the unit circle must contain the origin. However, the integrand $1/z$ and its indefinite integral $\mathrm{Ln}\,z$ are not analytic at the origin, and thus, Eq. 4.23 cannot be applied to evaluate the definite integral.

4.2.4 Cauchy's Integral Theorem

Let us express the integrand $f(z)$ and the differential dz in terms of their real and imaginary parts as

$$
\begin{cases}
f(z) = u(x, y) + iv(x, y) \\
dz = z'(t)dt = (x'(t) + iy'(t))dt = dx + i\,dy
\end{cases}
\tag{4.24}
$$

Then, the complex integral Eq. 4.22 for a closed contour is rewritten as

$$
\oint_C f(z)dz = \oint_C (u + iv)(dx + i\,dy)
$$

$$
= \oint_C (u\,dx - v\,dy) + i \oint_C (v\,dx + u\,dy)
\tag{4.25}
$$

Let us consider a region R consisting of all points interior to and on a simple closed contour C. If $f(z)$ is analytic in R, $f(z)$ is continuous and its derivative $f'(z)$ exists in R. Consequently, the functions u and v are continuous in R. Similarly, if $f'(z)$ is assumed to be continuous in R, u and v have continuous first-order partial derivatives in R. Hence, Green's theorem (Appendix B.1) is applicable and gives

$$
\begin{cases}
\oint_C (u\,dx - v\,dy) = \iint_R \left(-\dfrac{\partial v}{\partial x} - \dfrac{\partial u}{\partial y} \right) dxdy \\
\oint_C (v\,dx + u\,dy) = \iint_R \left(\dfrac{\partial u}{\partial x} - \dfrac{\partial v}{\partial y} \right) dxdy
\end{cases}
\tag{4.26}
$$

both of which vanish by virtue of the Cauchy–Riemann equations, and it follows that

$$
\oint_C f(z)dz = 0
\tag{4.27}
$$

The derivation of Eq. 4.27 given here is intuitive but requires the condition of continuity on $f'(z)$, which indeed can be removed. The proof without the condition that $f'(z)$ is continuous is more complicated and is put off to Appendix B.2. Here, let us state a universal version of the integral theorem that is fundamental throughout complex analysis.

Theorem 4.1 (Cauchy's integral theorem) *If $f(z)$ is analytic at all points interior to and on a simple closed contour C, then*

$$\oint_C f(z)dz = 0 \tag{4.28}$$

If the domain D of $f(z)$ is simply connected, then, by definition, the interior of any simple closed contour C, contained in D, is contained in D. Hence, Cauchy's integral theorem can be extended to a simply connected domain as follows.

Theorem 4.2 (Cauchy's integral theorem for a simply connected domain) *If $f(z)$ is analytic in a simply connected domain D, then*

$$\oint_C f(z)dz = 0 \tag{4.29}$$

for every simple closed contour C in D.

Corollary 4.1 (Independence of contour) *If $f(z)$ is analytic in a simply connected domain D, then the integral of $f(z)$ is independent of the contour in D.*

Proof Let us consider two contours C_1 and C_2 in D from any fixed point z_1 to any fixed point z_2 in D without further common points. By Cauchy's integral theorem, the integral from z_1 over C_1 to z_2 and over $-C_2$ back to z_1 is zero, since the path is a simple closed contour, that is,

$$\int_{C_1} f(z)dz + \int_{-C_2} f(z)dz = 0$$

From Eq. 4.18, reversing the contour $-C_2$ from z_1 to z_2 introduces a minus sign and it follows that

$$\int_{C_1} f(z)dz - \int_{C_2} f(z)dz = 0$$

which means that the integrals of $f(z)$ over C_1 and C_2 are equal and proves the corollary.   □

Example 4.9 Let us revisit Examples 4.3 and 4.4. Since $f(z) = \operatorname{Re} z = x$ is not analytic, the integral of $\operatorname{Re} z$ depends on the contours as seen in Example 4.3. On the other hand, $f(z) = z$ is analytic in the entire complex plane, which is certainly simply connected, and thus, the integral of z is independent of the contour as seen in Example 4.4.

Fig. 4.8 Contours C and C_j
connected by cuts L_j in a
multiply connected domain

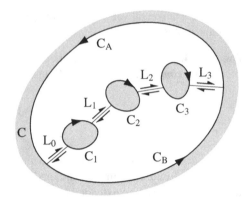

A further extension of Cauchy's integral theorem to multiply connected domains
can be made as follows.

Theorem 4.3 (Cauchy's integral theorem in a multiply connected domain) *Let C be a
simple closed contour oriented in the counterclockwise direction and $C_1, C_2, \ldots, C_n$
be simple closed contours oriented in the clockwise direction. The contours C_j ($j =
1, 2, \ldots, n$) are interior to C and their interiors have no points in common. If $f(z)$
is analytic in a domain that contains all points interior to and on C except for the
points interior to each C_j, then*

$$\oint_C f(z)dz + \sum_{j=1}^{n} \oint_{C_j} f(z)dz = 0 \qquad (4.30)$$

Proof Let us introduce a cut L_0 to connect the outer contour C to the inner contour
C_1. Additional cuts L_j individually connect C_j to C_{j+1} for $j = 1, 2, \ldots, n-1$ and
another cut L_n connects C_n to the outer contour C, as shown in Fig. 4.8 for the case
$n = 3$.

Eventually, two simple closed contours C_A and C_B are formed, each of which
consists of cuts L_j or $-L_j$ and pieces of C and C_j. By Cauchy's integral theorem,
the integrals over the contour of C_A and over the contour of C_B are both zero, and
the sum of those integrals is zero. In this sum, since the integrals over the cuts L_j and
$-L_j$ are in opposite directions and cancel, only the integrals over C and C_j remain,
which proves the theorem.   □

When there is a simple closed contour C_1 oriented in the counterclockwise direction interior to C, from Eq. 4.30, it follows that

$$\oint_C f(z)dz + \oint_{-C_1} f(z)dz = 0 \qquad (4.31)$$

which leads to an important consequence of the integral theorem as follows.

Corollary 4.2 (Principle of deformation of contours) *Let C and C_1 be simple closed contours oriented in the counterclockwise direction, where C_1 is interior to C. If $f(z)$ is analytic in the closed region consisting of those contours and all points between them, then*

$$\oint_C f(z)dz = \oint_{C_1} f(z)dz \qquad (4.32)$$

This corollary implies that if the contour is continuously deformed into another, passing through points where f is analytic, the value of the definite integral of f over C is invariant.

Example 4.10 Let us evaluate the definite integral of $f(z) = (z - z_0)^m$ over a simple closed contour C in the counterclockwise direction, where m is an integer and z_0 is a fixed point interior to C. By the principle of deformation of contours, the integral is equivalent to the one over the circle C_1 of radius r with its center at z_0 that lies interior to C.

The contour C_1 is given by $z(\theta) = z_0 + re^{i\theta}$ for $0 \le \theta \le 2\pi$, and it follows that

$$\begin{cases} f(z(\theta)) = (z - z_0)^m = r^m e^{im\theta} \\ z'(\theta) = ire^{i\theta} \end{cases}$$

The definite integral along C is obtained as

$$\oint_C (z - z_0)^m dz = \oint_{C_1} (z - z_0)^m dz$$

$$= \int_0^{2\pi} r^m e^{im\theta} ire^{i\theta} d\theta = ir^{m+1} \int_0^{2\pi} e^{i(m+1)\theta} d\theta$$

$$= ir^{m+1} \int_0^{2\pi} \cos(m+1)\theta d\theta - r^{m+1} \int_0^{2\pi} \sin(m+1)\theta d\theta$$

where Euler's formula is used.

When $m = -1$, $r^{m+1} = 1$, $\cos(m+1)\theta = 1$, and $\sin(m+1)\theta = 0$, and the integral becomes $2\pi i$. When $m \ne -1$, each of the two integrals is zero. Hence, the desired result is

$$\begin{cases} \oint_C \dfrac{1}{z - z_0} dz = 2\pi i \\[4mm] \oint_C (z - z_0)^m dz = 0 \end{cases} \qquad (4.33)$$

where m is any integer except $m = -1$.

4.2.5 Cauchy's Integral Formula

One of the most important results in complex analysis is Cauchy's integral formula. Besides the evaluation of complex integrals, it plays an important role in developing a theory on the derivatives of analytic functions. Furthermore, the formula becomes a mathematical basis of the numerical scheme to be developed in Sect. 4.3.

Theorem 4.4 (Cauchy's integral formula) *If $f(z)$ is analytic everywhere within and on a simple closed contour C oriented in the counterclockwise direction and z_0 is any point interior to C, then*

$$f(z_0) = \frac{1}{2\pi i} \oint_C \frac{f(z)}{z - z_0} dz \qquad (4.34)$$

Proof Since $f(z)$ is analytic, it is continuous at z_0; thus, for a given positive number ε, there exists some positive number δ such that

$$|f(z) - f(z_0)| < \varepsilon \quad \text{whenever} \quad |z - z_0| < \delta$$

Then, it is possible to choose a positive number ρ less than δ such that

$$|f(z) - f(z_0)| < \varepsilon \quad \text{whenever} \quad |z - z_0| = \rho$$

and such that a circle C_ρ of radius ρ with its center at z_0 lies interior to C as shown in Fig. 4.9. It follows the inequality

$$\frac{|f(z) - f(z_0)|}{|z - z_0|} < \frac{\varepsilon}{\rho} \qquad (4.35)$$

at each point of C_ρ.

Fig. 4.9 A simple closed contour C containing a circular contour C_ρ of radius ρ with its center at z_0

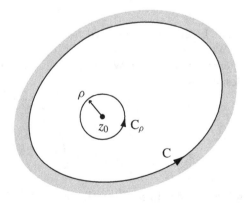

The integrand $f(z)/(z - z_0)$ is analytic in the closed region consisting of the contours C and C_ρ and all points between them; thus, from the principle of deformation of contours, it follows that

$$\oint_C \frac{f(z)}{z - z_0} dz = \oint_{C_\rho} \frac{f(z)}{z - z_0} dz \qquad (4.36)$$

where C_ρ is oriented in the counterclockwise direction. From Eq. 4.33, $f(z_0)$ can be written as

$$f(z_0) = \frac{f(z_0)}{2\pi i} \oint_{C_\rho} \frac{1}{z - z_0} dz = \frac{1}{2\pi i} \oint_{C_\rho} \frac{f(z_0)}{z - z_0} dz \qquad (4.37)$$

Substituting Eqs. 4.36 and 4.37 into Eq. 4.34 yields

$$\oint_{C_\rho} \frac{f(z_0)}{z - z_0} dz = \oint_{C_\rho} \frac{f(z)}{z - z_0} dz \qquad (4.38)$$

To examine whether Eq. 4.38 is valid, let us evaluate

$$\left| \oint_{C_\rho} \frac{f(z)}{z - z_0} dz - \oint_{C_\rho} \frac{f(z_0)}{z - z_0} dz \right| = \left| \oint_{C_\rho} \frac{f(z) - f(z_0)}{z - z_0} dz \right|$$

$$\leq \oint_{C_\rho} \frac{|f(z) - f(z_0)|}{|z - z_0|} |dz|$$

$$< \frac{\varepsilon}{\rho} 2\pi\rho = 2\pi\varepsilon$$

where Eq. 4.35 and the *ML* inequality (Eq. 4.21) are used. Since ε can be arbitrarily small, the left-hand side must be equal to zero. Hence, Eq. 4.38 is valid and the Cauchy's integral formula is proved.   □

Cauchy's integral formula states that the value of an analytic function $f(z)$ can be represented by a certain contour integral. Likewise, the nth derivative, $f^{(n)}(z)$, is represented by an analogous contour integral, and it can be shown that analytic functions have derivatives of all orders, which completely differs from real-variable calculus.

Theorem 4.5 (Cauchy's integral formula for derivatives) *If $f(z)$ is analytic everywhere within and on a simple closed contour C oriented in the counterclockwise direction and z_0 is any point interior to C, then for any integer $n \geq 0$*

$$f^{(n)}(z_0) = \frac{n!}{2\pi i} \oint_C \frac{f(z)}{(z - z_0)^{n+1}} dz \qquad (4.39)$$

Proof For $n = 0$, $f^{(0)}(z) = f(z)$, and Eq. 4.39 reduces to Eq. 4.34; hence, the theorem for the case $n = 0$ is proved.

For $n = 1$, the derivative is given by

$$f'(z_0) = \lim_{\Delta z \to 0} \frac{f(z_0 + \Delta z) - f(z_0)}{\Delta z}$$

By Cauchy's integral formula, it follows that

$$\frac{f(z_0 + \Delta z) - f(z_0)}{\Delta z} = \frac{1}{2\pi i \Delta z} \left[\oint_C \frac{f(z)}{z - (z_0 + \Delta z)} dz - \oint_C \frac{f(z)}{z - z_0} dz \right]$$

$$= \frac{1}{2\pi i} \oint_C \frac{f(z)}{(z - z_0 - \Delta z)(z - z_0)} dz$$

$$= \frac{1}{2\pi i} \oint_C \frac{f(z)}{(z - z_0)^2} dz + \frac{1}{2\pi i} \oint_C \frac{\Delta z f(z)}{(z - z_0 - \Delta z)(z - z_0)^2} dz$$

where, to validate the theorem for the case $n = 1$, the last term must be zero as $\Delta z \to 0$.

Since $f(z)$ is analytic, it is continuous on C and bounded in absolute value; $|f(z)| \le M$. Let d be the smallest distance from z_0 to C, then for all z on C

$$\frac{1}{|z - z_0|^2} \le \frac{1}{d^2} \tag{4.40}$$

and

$$d \le |z - z_0| = |z - z_0 - \Delta z + \Delta z| \le |z - z_0 - \Delta z| + |\Delta z|$$

which, in choosing $|\Delta z|$ so small that $d - |\Delta z| > 0$, yields

$$\frac{1}{|z - z_0 - \Delta z|} \le \frac{1}{d - |\Delta z|} \tag{4.41}$$

Let L be the length of C; then from Eqs. 4.40 and 4.41 and the *ML* inequality, it follows that

$$\left| \oint_C \frac{\Delta z f(z)}{(z - z_0 - \Delta z)(z - z_0)^2} dz \right| \le \frac{|\Delta z| ML}{(d - |\Delta z|) d^2}$$

which approaches zero as $\Delta z \to 0$, and consequently

$$\lim_{\Delta z \to 0} \frac{f(z_0 + \Delta z) - f(z_0)}{\Delta z} = \frac{1}{2\pi i} \oint_C \frac{f(z)}{(z - z_0)^2} dz$$

which proves the theorem for the case $n = 1$.

The same argument, with $f(z)$ replaced by $f'(z)$, can be applied to the analytic function $f'(z)$, which proves that the derivative $f''(z_0)$ is represented by Eq. 4.39 for $n = 2$. In a similar manner, the results for $n = 3, 4, \ldots$ are established, and the general formula Eq. 4.39 follows by induction. $\qquad\qquad\qquad\qquad\qquad\qquad\qquad\square$

If $f(z)$ is analytic at a point z_0, there exists a circle about z_0 within and on which $f(z)$ is analytic. From Eq. 4.39, $f^{(n)}(z_0)$ exists at each point interior to the circle, which implies that $f^{(n-1)}(z)$ is analytic at z_0. This leads to the following corollary.

Corollary 4.3 (Existence of derivatives of all orders) *If $f(z)$ is analytic at a point, then its derivatives $f^{(n)}(z)$ of all orders exist and therefore are analytic at that point.*

Example 4.11 Let z_0 be a fixed point interior to a simple closed contour C oriented in the counterclockwise direction. When $f(z) = 1$, Cauchy's integral formula Eq. 4.34 shows

$$\oint_C \frac{1}{z - z_0} dz = 2\pi i$$

In contrast, since $f^{(n)}(z) = 0$ for $n = 1, 2, \ldots$, Cauchy's integral formula Eq. 4.39 shows

$$\oint_C \frac{1}{(z - z_0)^{n+1}} dz = 0$$

Note that these results are consistent with Eq. 4.33 in Example 4.10.

In a similar way to how Cauchy's integral theorem is extended to multiply connected domains (Theorem 4.3), Cauchy's integral formula can be extended to multiply connected domains by replacing the simple closed contour C in the formula by the boundary of a multiply connected domain.

Corollary 4.4 (Cauchy's integral formula in a multiply connected domain) *Let C be a simple closed contour oriented in the counterclockwise direction and $C_1, C_2, \ldots, C_n$ be simple closed contours oriented in the clockwise direction. The contours C_j ($j = 1, 2, \ldots, n$) are interior to C and their interiors have no points in common. If $f(z)$ is analytic in a domain that contains all points interior to and on C except for the points interior to each C_j, then*

$$f(z_0) = \frac{1}{2\pi i} \left[\oint_C \frac{f(z)}{z - z_0} dz + \sum_{j=1}^{n} \oint_{C_j} \frac{f(z)}{z - z_0} dz \right] \qquad (4.42)$$

where z_0 is any point in the domain.

Suppose, for instance, $f(z)$ is analytic on C, C_1, C_2, and C_3 and within the domain bounded by these boundaries, as shown in Fig. 4.10. With appropriate cuts between

Fig. 4.10 Contours C and C_j with cuts, resulting in a simply connected domain

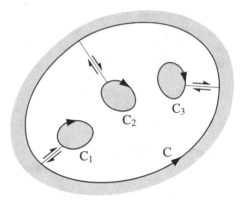

C and C_j for $j = 1, 2$, and 3, the domain becomes simply connected, which verify the following formula:

$$f(z_0) = \frac{1}{2\pi i} \left[\oint_C \frac{f(z)}{z - z_0} dz + \oint_{C_1} \frac{f(z)}{z - z_0} dz + \oint_{C_2} \frac{f(z)}{z - z_0} dz + \oint_{C_3} \frac{f(z)}{z - z_0} dz \right]$$

(4.43)

Note that the orientation of C is in the counterclockwise direction, while the orientation of C_j is in the clockwise direction, and that integrals over the cuts cancel because the integrations are in both directions.

4.3 Complex Variable Boundary Element Method

In practice, there exist many boundary value problems that cannot be solved by analytical methods or the solutions of which are so difficult that numerical approaches are required. Numerical analysis is indispensable to practical engineering. The most pertinent numerical scheme in complex analysis is the complex variable boundary element method (CVBEM).

4.3.1 Mathematical Preliminaries of the CVBEM

The CVBEM owes its elegance, the computational accuracy and efficiency, to Cauchy's integral formula. Because of its sound mathematical foundations, the CVBEM can be interpreted as a semi-analytical scheme with versatility.

For the complex potential Ω that is analytic everywhere within and on a simple closed boundary Γ, Cauchy's integral formula holds for any point z interior to Γ

$$\Omega(z) = \frac{1}{2\pi i} \oint_\Gamma \frac{\Omega(\zeta)}{\zeta - z} d\zeta$$

(4.44)

where ζ is on the boundary Γ and the contour integration is taken in the counter-clockwise direction.

Proposition 4.1 (Interior solutions of complex potential) *The value of complex potential Ω at any point interior to a simple closed boundary Γ is completely determined by the values of Ω on Γ.*

Cauchy's integral formula is a powerful tool to obtain interior solutions $\Omega(z)$, provided that the contour integral in Eq. 4.44 can be evaluated and that the boundary values $\Omega(\zeta)$ are all known along Γ. In general, however, the integral cannot be manipulated analytically, except for some special cases. In addition, as mentioned in Sect. 4.1.3, only half of the boundary conditions are prescribed in well-posed boundary value problems, and the remaining half must be determined. These two difficulties limit the applicability of Cauchy's integral formula in its original form of Eq. 4.44.

In the process of the CVBEM, therefore, determination of the complete boundary values $\Omega(\zeta)$ needs to precede evaluation of interior solutions $\Omega(z)$. A set of equations required to determine the unprescribed boundary values can be obtained by allocating the point z on an appropriate number of points on the boundary. This procedure, as well as evaluation of integrals, requires boundary discretization and interpolation. Once the boundary unknowns are determined, Cauchy's integral formula can be used to compute interior solutions $\Omega(z)$ at any location z interior to the boundary Γ.

4.3.2 Discretization

To overcome the difficulty in integration, the integral along the boundary is approximated by the summation of piecewise contour integrals. The boundary Γ is discretized into n_b straight-line elements $\Delta\Gamma_j$; that is, Γ is approximated with $\Delta\Gamma_1 \cup \Delta\Gamma_2 \cup \cdots \cup \Delta\Gamma_{n_b}$. The approximate expression for Eq. 4.44 is obtained as

$$\tilde{\Omega}(z) = \frac{1}{2\pi i} \sum_{j=1}^{n_b} \int_{\Delta\Gamma_j} \frac{\hat{\Omega}(\zeta)}{\zeta - z} d\zeta \tag{4.45}$$

where $\tilde{\Omega}(z)$ is the approximate value of $\Omega(z)$ and $\hat{\Omega}(\zeta)$ is the trial function along the boundary elements $\Delta\Gamma_j$. While the numbering direction is arbitrary, the positive direction is chosen in accordance with the convention regarding Cauchy's integral formula.

To define a profile of $\hat{\Omega}(\zeta)$ along the n_b boundary elements, a finite number of nodes, or nodal points, are distributed on Γ, where the boundary values are associated. With the nodal values Ω_j at the nodal points ζ_j, the overall profile of the trial function $\hat{\Omega}(\zeta)$ is given by

$$\hat{\Omega}(\zeta) = \sum_{j=1}^{n_b} N_j(\zeta)\Omega_j \tag{4.46}$$

Table 4.1 Summary of the CVBEM variables

Variable	Description	Related equation
$\tilde{\Omega}(z)$	Approximate values of $\Omega(z)$	Given by Eq. 4.45
$\hat{\Omega}(\zeta)$	Trial functions to compute $\tilde{\Omega}(z)$	Used in Eq. 4.45, given by Eq. 4.46
Ω_j	Nodal values to evaluate $\hat{\Omega}(\zeta)$	Used in Eq. 4.46

where $N_j(\zeta)$ is the interpolation function on the boundary element $\Delta\Gamma_j$. In the following sections, two types of interpolation schemes are considered.

Table 4.1 summarizes the variables in the CVBEM, which are used for discretization and interpolation.

4.3.2.1 Constant Interpolation

The simplest interpolation scheme is piecewise constant interpolation, which assigns individual constant values over each of the boundary elements. Figure 4.11 shows a sample discretization of the boundary Γ (a dashed line) into n_b boundary elements $\Delta\Gamma_j$ (solid line segments). The nodal point ζ_j is located at the middle of the element. For the n_b-boundary discretization, n_b nodes are required.

Given the complex potential Ω_j at the nodal point ζ_j, constant interpolation gives the trial function $\hat{\Omega}(\zeta)$ on $\Delta\Gamma_j$ as

$$\hat{\Omega}(\zeta) = \Omega_j = N_j(\zeta)\Omega_j \tag{4.47}$$

The interpolation function $N_j(\zeta)$ is equal to 1 over each element $\Delta\Gamma_j$, as shown in Fig. 4.12. Constant interpolation relies on the assumption that the rate of change along the boundary element is zero and allows sudden changes of boundary values at the element intersections.

Fig. 4.11 Discretization of boundary Γ (*dashed line*) into constant elements $\Delta\Gamma_j$ (*solid line* segments)

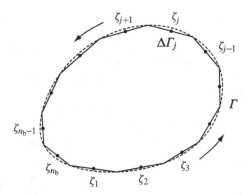

Fig. 4.12 Constant interpolation function $N_j(\zeta) = 1$ on a boundary element $\Delta\Gamma_j$

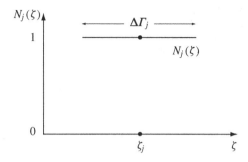

Fig. 4.13 Interpolation with constant elements; interpolated (*solid lines*) and exact (*dashed line*) profiles

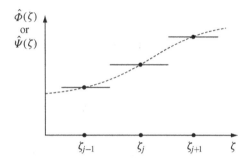

For the overall interpolation, Eq. 4.46, the constant interpolation function $N_j(\zeta)$ is defined by

$$N_j(\zeta) = \begin{cases} 1 & (\zeta \in \Delta\Gamma_j) \\ 0 & (\zeta \notin \Delta\Gamma_j) \end{cases} \tag{4.48}$$

The profile of $\hat{\Omega}(\zeta) = \hat{\Phi}(\zeta) + i\hat{\Psi}(\zeta)$ with constant interpolation suffers discontinuities at the element intersections, as exemplified in Fig. 4.13 showing the profile of $\hat{\Phi}(\zeta)$ or $\hat{\Psi}(\zeta)$. Despite the simplicity of constant interpolation, the discontinuities of $\hat{\Omega}(\zeta)$ discourage us from its application.

4.3.2.2 Linear Interpolation

Linear interpolation overcomes the problem of discontinuities at the element intersections. The nodal points are allocated at the edges of linear elements, as shown in Fig. 4.14. The jth boundary element $\Delta\Gamma_j$ is defined by the nodal points ζ_j and ζ_{j+1}, where the nodal values are specified, and $\hat{\Omega}(\zeta)$ is linearly interpolated in between. Note that ζ_{n_b+1} coincides with ζ_1 and the element $\Delta\Gamma_{n_b}$ is defined by ζ_{n_b} and ζ_1. As is the case with constant interpolation, n_b nodes are required for the n_b-boundary discretization.

Fig. 4.14 Discretization of
boundary Γ (*dashed line*)
into linear elements $\Delta\Gamma_j$
(*solid line* segments)

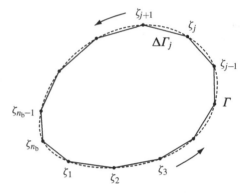

Given the complex potentials Ω_j and Ω_{j+1} at the nodal points ζ_j and ζ_{j+1}, respectively, linear interpolation gives $\hat{\Omega}(\zeta)$ on $\Delta\Gamma_j$ as

$$
\hat{\Omega}(\zeta) = \frac{\zeta_{j+1} - \zeta}{\zeta_{j+1} - \zeta_j}\Omega_j + \frac{\zeta - \zeta_j}{\zeta_{j+1} - \zeta_j}\Omega_{j+1}
$$
$$
= N_j(\zeta)\Omega_j + N_{j+1}(\zeta)\Omega_{j+1} \tag{4.49}
$$

The profiles of interpolation functions $N_j(\zeta)$ and $N_{j+1}(\zeta)$ are shown in Fig. 4.15; $N_j(\zeta)$ is 1 at ζ_j and 0 at ζ_{j+1}, similarly, $N_{j+1}(\zeta)$ is 0 at ζ_j and 1 at ζ_{j+1}. Linear interpolation relies on the assumption that the rate of change between the nodal points is constant and can be calculated from the corresponding nodal values using a simple slope formula.

For the overall interpolation, Eq. 4.46, the linear interpolation function $N_j(\zeta)$ is defined by

$$
N_j(\zeta) = \begin{cases}
\dfrac{\zeta - \zeta_{j-1}}{\zeta_j - \zeta_{j-1}} & (\zeta \in \Delta\Gamma_{j-1}) \\[2ex]
\dfrac{\zeta_{j+1} - \zeta}{\zeta_{j+1} - \zeta_j} & (\zeta \in \Delta\Gamma_j) \\[2ex]
0 & (\zeta \notin \Delta\Gamma_{j-1} \cup \Delta\Gamma_j)
\end{cases} \tag{4.50}
$$

Fig. 4.15 Linear
interpolation functions
$N_j(\zeta)$ and $N_{j+1}(\zeta)$ on a
boundary element $\Delta\Gamma_j$

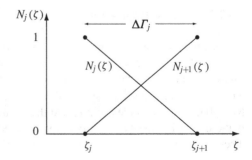

Fig. 4.16 Interpolation with linear elements; interpolated (*solid lines*) and exact (*dashed line*) profiles

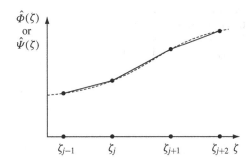

Linear interpolation provides a continuous profile of $\hat{\Omega}(\zeta) = \hat{\Phi}(\zeta) + i\hat{\Psi}(\zeta)$ at the element intersections. Figure 4.16 shows the profile of $\hat{\Phi}(\zeta)$ or $\hat{\Psi}(\zeta)$, which is linearly interpolated between the nodal points.

4.3.3 Cauchy's Integral Formula in Discretized Form

The discretized form of Cauchy's integral formula with linear interpolation is derived. Substituting Eq. 4.49 with Eq. 4.50 into Eq. 4.45 yields

$$
\tilde{\Omega}(z) = \frac{1}{2\pi i} \sum_{j=1}^{n_b} \int_{\Delta\Gamma_j} \frac{\hat{\Omega}(\zeta)}{\zeta - z} d\zeta
$$

$$
= \frac{1}{2\pi i} \sum_{j=1}^{n_b} \int_{\Delta\Gamma_j} \frac{(\zeta - \zeta_j)\Omega_{j+1} + (\zeta_{j+1} - \zeta)\Omega_j}{(\zeta_{j+1} - \zeta_j)(\zeta - z)} d\zeta
$$

$$
= \frac{1}{2\pi i} \sum_{j=1}^{n_b} \left[\frac{\zeta_{j+1}\Omega_j - \zeta_j\Omega_{j+1}}{\zeta_{j+1} - \zeta_j} \int_{\Delta\Gamma_j} \frac{d\zeta}{\zeta - z} + \frac{\Omega_{j+1} - \Omega_j}{\zeta_{j+1} - \zeta_j} \int_{\Delta\Gamma_j} \frac{\zeta d\zeta}{\zeta - z} \right] \quad (4.51)
$$

The first integral on the right-hand side is evaluated as

$$
\int_{\Delta\Gamma_j} \frac{d\zeta}{\zeta - z} = \ln \frac{\zeta_{j+1} - z}{\zeta_j - z} \quad (4.52)
$$

and the second integral as

$$
\int_{\Delta\Gamma_j} \frac{\zeta d\zeta}{\zeta - z} = \int_{\Delta\Gamma_j} \left(1 + \frac{z}{\zeta - z}\right) d\zeta
$$

$$
= \zeta_{j+1} - \zeta_j + z \ln \frac{\zeta_{j+1} - z}{\zeta_j - z} \quad (4.53)
$$

Fig. 4.17 Angle $\alpha_j(z)$
between the line segments
joining the nodal points ζ_j
and ζ_{j+1} to the interior point
z

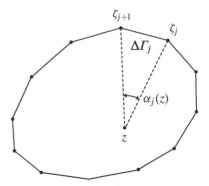

in which

$$\ln \frac{\zeta_{j+1} - z}{\zeta_j - z} = \ln \left| \frac{\zeta_{j+1} - z}{\zeta_j - z} \right| + i\alpha_j(z) \tag{4.54}$$

where $\alpha_j(z)$ is the angle between the line segments joining the nodal points ζ_j and ζ_{j+1} to the interior point z, as shown in Fig. 4.17.

From these integral evaluations, the integral over the boundary element $\Delta\Gamma_j$ is obtained as

$$\int_{\Delta\Gamma_j} \frac{\hat{\Omega}(\zeta)}{\zeta - z} d\zeta = \frac{\zeta_{j+1}\Omega_j - \zeta_j\Omega_{j+1}}{\zeta_{j+1} - \zeta_j} \ln \frac{\zeta_{j+1} - z}{\zeta_j - z}$$

$$+ \frac{\Omega_{j+1} - \Omega_j}{\zeta_{j+1} - \zeta_j} \left(\zeta_{j+1} - \zeta_j + z \ln \frac{\zeta_{j+1} - z}{\zeta_j - z} \right)$$

$$= \frac{(z - \zeta_j)\Omega_{j+1} - (z - \zeta_{j+1})\Omega_j}{\zeta_{j+1} - \zeta_j} \ln \frac{\zeta_{j+1} - z}{\zeta_j - z} + \Omega_{j+1} - \Omega_j \tag{4.55}$$

and finally, the discretized form of Cauchy's integral formula with linear elements becomes

$$\tilde{\Omega}(z) = \frac{1}{2\pi i} \sum_{j=1}^{n_b} \frac{(z - \zeta_j)\Omega_{j+1} - (z - \zeta_{j+1})\Omega_j}{\zeta_{j+1} - \zeta_j} \ln \frac{\zeta_{j+1} - z}{\zeta_j - z} \tag{4.56}$$

where $\sum_{j=1}^{n_b}(\Omega_{j+1} - \Omega_j) = 0$ is used. Equation 4.56 lays the mathematical foundations for the CVBEM.

4.3.4 Nodal Equation

For the purpose of determining the unprescribed boundary values (the stream function Ψ on the Dirichlet-type boundary Γ_P and the velocity potential Φ on the Neumann-type boundary Γ_S), let us derive a nodal equation by considering the limiting value of $\tilde{\Omega}(z)$ (Eq. 4.45) as the interior point z approaches a nodal point ζ_k, as shown in Fig. 4.18.

Equation 4.45 can be rewritten as

$$2\pi i \tilde{\Omega}(z) = \int\limits_{\Delta\Gamma_{k-1}} \frac{\hat{\Omega}(\zeta)}{\zeta - z}d\zeta + \int\limits_{\Delta\Gamma_k} \frac{\hat{\Omega}(\zeta)}{\zeta - z}d\zeta + \sum_{\substack{j=1 \\ j\neq k-1 \\ j\neq k}}^{n_b} \int\limits_{\Delta\Gamma_j} \frac{\hat{\Omega}(\zeta)}{\zeta - z}d\zeta \qquad (4.57)$$

From Eq. 4.55, the limiting value of the first integral as $z \to \zeta_k$ becomes

$$\lim_{z\to\zeta_k} \int\limits_{\Delta\Gamma_{k-1}} \frac{\hat{\Omega}(\zeta)}{\zeta - z}d\zeta = \Omega_k \lim_{z\to\zeta_k} \ln \frac{\zeta_k - z}{\zeta_{k-1} - z} + \Omega_k - \Omega_{k-1} \qquad (4.58)$$

In a similar way, the limiting value of the second integral becomes

$$\lim_{z\to\zeta_k} \int\limits_{\Delta\Gamma_k} \frac{\hat{\Omega}(\zeta)}{\zeta - z}d\zeta = \Omega_k \lim_{z\to\zeta_k} \ln \frac{\zeta_{k+1} - z}{\zeta_k - z} + \Omega_{k+1} - \Omega_k \qquad (4.59)$$

The summation of these limiting integrals as $z \to \zeta_k$ yields

$$\lim_{z\to\zeta_k} \left[\int\limits_{\Delta\Gamma_{k-1}} \frac{\hat{\Omega}(\zeta)}{\zeta - z}d\zeta + \int\limits_{\Delta\Gamma_k} \frac{\hat{\Omega}(\zeta)}{\zeta - z}d\zeta \right]$$

$$= \Omega_{k+1} - \Omega_{k-1} + \Omega_k \ln \frac{\zeta_{k+1} - \zeta_k}{\zeta_{k-1} - \zeta_k}$$

$$= \Omega_{k+1} - \Omega_{k-1} + \Omega_k \ln \left| \frac{\zeta_{k+1} - \zeta_k}{\zeta_{k-1} - \zeta_k} \right| + i(2\pi - \theta_k)\Omega_k \qquad (4.60)$$

where θ_k is the interior angle at the nodal point ζ_k, as shown in Fig. 4.18.

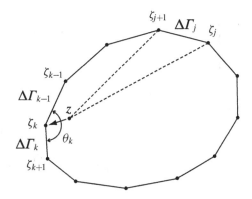

Fig. 4.18 Interior point z approaching a nodal point ζ_k at which the interior angle is θ_k

The third integral is also evaluated using Eq. 4.55, and finally, when $z \to \zeta_k$, Eq. 4.57 is rewritten as

$$
2\pi i \tilde{\Omega}(\zeta_k) = \Omega_k \ln \left| \frac{\zeta_{k+1} - \zeta_k}{\zeta_{k-1} - \zeta_k} \right| + i(2\pi - \theta_k)\Omega_k
$$
$$
+ \sum_{\substack{j=1 \\ j \neq k-1 \\ j \neq k}}^{n_b} \frac{(\zeta_k - \zeta_j)\Omega_{j+1} - (\zeta_k - \zeta_{j+1})\Omega_j}{\zeta_{j+1} - \zeta_j} \ln \frac{\zeta_{j+1} - \zeta_k}{\zeta_j - \zeta_k} \qquad (4.61)
$$

Equation 4.61 is the nodal equation that relates the approximate boundary value $\tilde{\Omega}(\zeta_k) = \tilde{\Phi}(\zeta_k) + i\tilde{\Psi}(\zeta_k)$ at the nodal point ζ_k with the nodal values Ω_j at the nodal points ζ_j. For the velocity potential, the imaginary part of Eq. 4.61 gives

$$
2\pi \tilde{\Phi}(\zeta_k) = \Psi_k \ln \left| \frac{\zeta_{k+1} - \zeta_k}{\zeta_{k-1} - \zeta_k} \right| + (2\pi - \theta_k)\Phi_k + \mathrm{Im}[G_k] \qquad (4.62)
$$

and for the stream function, the real part gives

$$
-2\pi \tilde{\Psi}(\zeta_k) = \Phi_k \ln \left| \frac{\zeta_{k+1} - \zeta_k}{\zeta_{k-1} - \zeta_k} \right| - (2\pi - \theta_k)\Psi_k + \mathrm{Re}[G_k] \qquad (4.63)
$$

where

$$
G_k = \sum_{\substack{j=1 \\ j \neq k-1 \\ j \neq k}}^{n_b} \frac{(\zeta_k - \zeta_j)\Omega_{j+1} - (\zeta_k - \zeta_{j+1})\Omega_j}{\zeta_{j+1} - \zeta_j} \ln \frac{\zeta_{j+1} - \zeta_k}{\zeta_j - \zeta_k} \qquad (4.64)
$$

The real and imaginary parts of G_k can be written as

$$
\begin{cases}
\mathrm{Re}[G_k] = \displaystyle\sum_{\substack{j=1 \\ j \neq k-1 \\ j \neq k}}^{n_b} \left(\mathrm{Re}[H_j]\Phi_{j+1} - \mathrm{Im}[H_j]\Psi_{j+1} - \mathrm{Re}[I_j]\Phi_j + \mathrm{Im}[I_j]\Psi_j \right) \\[2em]
\mathrm{Im}[G_k] = \displaystyle\sum_{\substack{j=1 \\ j \neq k-1 \\ j \neq k}}^{n_b} \left(\mathrm{Re}[H_j]\Psi_{j+1} + \mathrm{Im}[H_j]\Phi_{j+1} - \mathrm{Re}[I_j]\Psi_j - \mathrm{Im}[I_j]\Phi_j \right)
\end{cases}
$$
$$
(4.65)
$$

in which H_j and I_j are given by

$$
\begin{cases}
H_j = \dfrac{\zeta_k - \zeta_j}{\zeta_{j+1} - \zeta_j} \ln \dfrac{\zeta_{j+1} - \zeta_k}{\zeta_j - \zeta_k} \\[1.5em]
I_j = \dfrac{\zeta_k - \zeta_{j+1}}{\zeta_{j+1} - \zeta_j} \ln \dfrac{\zeta_{j+1} - \zeta_k}{\zeta_j - \zeta_k}
\end{cases}
\qquad (4.66)
$$

Note that all the coefficients of Φ_j and Ψ_j depend only on the geometry of the boundary and can be evaluated independently of boundary conditions.

4.4 Formulations of the CVBEM

The nodal equation is used along with the prescribed boundary values to determine unprescribed boundary values. There are two types of boundary values appearing in Eq. 4.61: approximate boundary values $\tilde{\Omega}(\zeta_j) = \tilde{\Phi}(\zeta_j) + i\tilde{\Psi}(\zeta_j)$ and nodal values $\Omega_j = \Phi_j + i\Psi_j$.

In formulating the CVBEM, $\tilde{\Omega}(\zeta_j)$ and/or Ω_j are equated with the prescribed boundary values $P_j + iS_j$. Depending on different combinations of equivalence, three types of formulations are provided. Table 4.2 summarizes the formulations, which are detailed in the following sections.

To assess their performances,[1] individual formulations are applied to the problem discussed in Example 3.13: flow around a cylindrical corner. As shown in Fig. 4.19, it is assumed that the stream function along the boundary DA_1A_2B and the velocity potential along the boundary BCD are prescribed.

According to the analytical solution Eq. 3.41, the boundary conditions are given as follows:

$$
\begin{cases}
S = 0 & \text{(along } DA_1A_2B) \\[2mm]
P = a(1 - y^2)\left[1 + \dfrac{0.25^2}{(1 + y^2)^2}\right] & \text{(along } BC) \\[2mm]
P = a(x^2 - 1)\left[1 + \dfrac{0.25^2}{(x^2 + 1)^2}\right] & \text{(along } CD)
\end{cases}
\tag{4.67}
$$

where $a = 1/0.984375$. The profiles of S along DA_1A_2B and P along BCD are shown in Fig. 4.19.

Table 4.2 Summary of the CVBEM formulations

Formulation	Equivalence	Dirichlet-type boundary	Neumann-type boundary
I	Known variable	$\tilde{\Phi}(\zeta_k) = P_k$	$\tilde{\Psi}(\zeta_k) = S_k$
		$\Phi_k = P_k$	$\Psi_k = S_k$
II	Unknown variable	$\Phi_k = P_k$	$\tilde{\Phi}(\zeta_k) = \Phi_k$
		$\tilde{\Psi}(\zeta_k) = \Psi_k$	$\Psi_k = S_k$
III	Dual variables	$\tilde{\Phi}(\zeta_k) = P_k$	$\tilde{\Phi}(\zeta_k) = \Phi_k$
		$\tilde{\Psi}(\zeta_k) = \Psi_k$	$\tilde{\Psi}(\zeta_k) = S_k$

---

[1] To exemplify the solution procedure of the CVBEM more clearly, a simpler problem is solved with Formulation III in Appendix E.1.

Fig. 4.19 Boundary
conditions of flow around a
cylindrical corner; S (*bold
solid lines*) and P (*bold
dashed lines*)

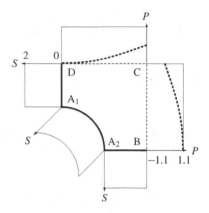

Table 4.3 Number of nodes used in the CVBEM computations

n_b	A_1A_2	A_2B	BC	CD	DA_1
16	4	2	4	4	2
32	8	4	8	8	4
48	12	6	12	12	6
64	16	8	16	16	8
80	20	10	20	20	10

To solve this problem, the CVBEM with various n_b values (from 16 to 80) is
applied. Table 4.3 summarizes the node distributions along the boundary for different
values of n_b.

4.4.1 Known-Variable Equivalence

In the known-variable equivalence (Formulation I), the nodal equation for the prescribed variable (known variable) is adopted and presumed to be equal to the prescribed boundary value and the nodal value of the same kind.

When the velocity potential P_k is prescribed at the nodal point ζ_k, the approximate
boundary value $\tilde{\Phi}(\zeta_k)$, given by the nodal equation for the velocity potential, and
the nodal value of the same kind Φ_k are equated with P_k:

$$\begin{cases} \tilde{\Phi}(\zeta_k) = P_k \\ \Phi_k = P_k \end{cases} \tag{4.68}$$

Substituting Eq. 4.68 into Eq. 4.62 gives

$$\theta_k P_k = \Psi_k \ln \left| \frac{\zeta_{k+1} - \zeta_k}{\zeta_{k-1} - \zeta_k} \right| + \text{Im}[G_k] \tag{4.69}$$

and Ψ_k is left unknown. Equation 4.63 is not utilized and the approximate boundary value of the stream function $\tilde{\Psi}(\zeta_k)$ may not be equal to the nodal value Ψ_k.

In a similar way, when the stream function S_k is prescribed at the nodal point ζ_k, the approximate boundary value $\tilde{\Psi}(\zeta_k)$, given by the nodal equation for the stream function, and the nodal value of the same kind Ψ_k are equated with S_k:

$$\begin{cases} \tilde{\Psi}(\zeta_k) = S_k \\ \Psi_k = S_k \end{cases} \tag{4.70}$$

Substituting Eq. 4.70 into Eq. 4.63 gives

$$-\theta_k S_k = \Phi_k \ln \left| \frac{\zeta_{k+1} - \zeta_k}{\zeta_{k-1} - \zeta_k} \right| + \text{Re}[G_k] \tag{4.71}$$

and Φ_k is left unknown. Equation 4.62 is not utilized and the approximate boundary value $\tilde{\Phi}(\zeta_k)$ may not be equal to the nodal value Φ_k.

In either of the boundary-condition cases, one of the nodal equations (Eq. 4.62 when P_k is given or Eq. 4.63 when S_k is given) is used. The result is a set of n_b equations for n_b nodal points, which yields unique solutions for n_b unprescribed boundary values.

4.4.1.1 Numerical Result

Formulation I applies the equivalence Eq. 4.68 along the boundary BCD and the equivalence Eq. 4.70 along the boundary DA_1A_2B. Figure 4.20 shows the approximate boundary values $\tilde{\Phi}(\zeta_k)$ and $\tilde{\Psi}(\zeta_k)$ obtained by the CVBEM with $n_b = 16$.

To assess the accuracy of the CVBEM, the computational errors $\tilde{E}_\Phi(\zeta_k)$ and $\tilde{E}_\Psi(\zeta_k)$ along the boundary, as defined below, are examined:

$$\begin{cases} \tilde{E}_\Phi(\zeta_k) = \tilde{\Phi}(\zeta_k) - \Phi(\zeta_k) \\ \tilde{E}_\Psi(\zeta_k) = \tilde{\Psi}(\zeta_k) - \Psi(\zeta_k) \end{cases} \tag{4.72}$$

where $\Phi(\zeta_k)$ and $\Psi(\zeta_k)$ are the exact solutions at the nodal point ζ_k, given by Eq. 3.41. The computational errors and exact solutions are also shown in Fig. 4.20.

Along the boundary BCD, Formulation I equates $\tilde{\Phi}(\zeta_k)$ with $P_k = \Phi(\zeta_k)$, and the CVBEM yields no error in $\tilde{\Phi}(\zeta_k)$, as seen in Fig. 4.20a. As for $\tilde{\Psi}(\zeta_k)$ along the boundary BCD, in contrast, a large error is observed, as shown in Fig. 4.20b. The reverse is true along the boundary DA_1A_2B; no error in $\tilde{\Psi}(\zeta_k)$ and a large amount of error in $\tilde{\Phi}(\zeta_k)$.

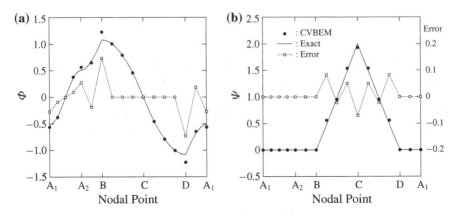

Fig. 4.20 Approximate boundary values, $\tilde{\Phi}(\zeta_k)$ and $\tilde{\Psi}(\zeta_k)$ (*black dots*), exact solutions, $\Phi(\zeta_k)$ and $\Psi(\zeta_k)$ (*solid lines*), and computational errors, $\tilde{E}_\Phi(\zeta_k)$ and $\tilde{E}_\Psi(\zeta_k)$ (*open squares*) for flow around a cylindrical corner using Formulation I with $n_b = 16$. **a** Velocity potential. **b** Stream function

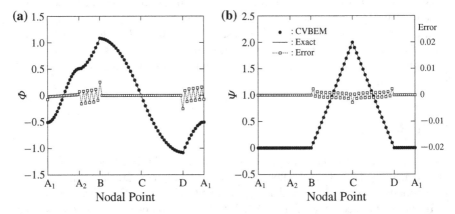

Fig. 4.21 Approximate boundary values, $\tilde{\Phi}(\zeta_k)$ and $\tilde{\Psi}(\zeta_k)$ (*black dots*), exact solutions, $\Phi(\zeta_k)$ and $\Psi(\zeta_k)$ (*solid lines*), and computational errors, $\tilde{E}_\Phi(\zeta_k)$ and $\tilde{E}_\Psi(\zeta_k)$ (*open squares*) for flow around a cylindrical corner by using Formulation I with $n_b = 80$. **a** Velocity potential. **b** Stream function

It is expected that the CVBEM results approach the exact solution as n_b increases. Figure 4.21 shows the solutions and errors with $n_b = 80$. Note that the scale for errors is different (10 times magnified) from Fig. 4.20. A significant reduction in errors is observed.

In either case ($n_b = 16$ or 80), severe oscillation is observed in the error profiles. Forcing the approximate boundary values and nodal values to be equal to the prescribed boundary values yields such errors in unprescribed boundary values.

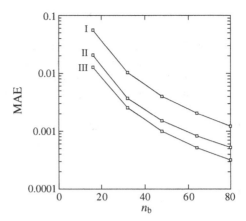

Fig. 4.22 Mean absolute error for different values of n_b; I for Formulation I, II for Formulation II, and III for Formulation III

The mean absolute error (MAE) in the complex potential is given by

$$\text{MAE} = \sum_{k=1}^{n_b} |\tilde{\Omega}(\zeta_k) - \Omega(\zeta_k)|/n_b \tag{4.73}$$

where $\tilde{\Omega}(\zeta_k) = \tilde{\Phi}(\zeta_k) + i\tilde{\Psi}(\zeta_k)$ and $\Omega(\zeta_k) = \Phi(\zeta_k) + i\Psi(\zeta_k)$ are the approximate and exact boundary solutions, respectively. Figure 4.22 summarizes the MAE characteristics with respect to n_b. As n_b increases, the MAE decreases; the MAE of 5.59×10^{-2} for $n_b = 16$ is reduced to 1.03×10^{-2} by doubling n_b, which is further reduced to 1.22×10^{-3} for $n_b = 80$. The solution with 80 nodes is 46 times more accurate than the one with 16 nodes.

It should be noted that these error estimations are based on the analytical (exact) solution of the problem, which is usually not available. In actual applications, the reference values which can be used in error estimations are the prescribed boundary values.

The major drawback of the known-variable equivalence lies in the fact that the prescribed boundary values are used in the equivalence equation and are of no use for error estimations. The true values of unprescribed boundary values are not known, and there is no way to check whether the boundary solutions are accurate enough for practical purposes.

4.4.2 Unknown-Variable Equivalence

In the unknown-variable equivalence (Formulation II), the nodal equation for the unprescribed variable (unknown variable) is adopted and presumed to be equal to the nodal value of the same kind. The nodal equation for the prescribed variable is not

used, and the prescribed boundary value is presumed to be equal to the corresponding
nodal value.

When the velocity potential P_k is prescribed at the nodal point ζ_k, the approximate boundary value $\tilde{\Psi}(\zeta_k)$, given by the nodal equation for the stream function
(a counterpart of the velocity potential), is equated with the nodal value Ψ_k. The
approximate boundary value $\tilde{\Phi}(\zeta_k)$ of the prescribed variable is not forced to be
equal to the prescribed boundary value P_k; instead, the nodal value Φ_k is equated
with P_k. Then, it follows that

$$
\begin{cases}
\Phi_k = P_k \\
\tilde{\Psi}(\zeta_k) = \Psi_k
\end{cases}
\tag{4.74}
$$

Substituting Eq. 4.74 into Eq. 4.63 gives

$$
-\theta_k \Psi_k = P_k \ln\left|\frac{\zeta_{k+1} - \zeta_k}{\zeta_{k-1} - \zeta_k}\right| + \mathrm{Re}[G_k]
\tag{4.75}
$$

and Ψ_k is left unknown.

In a similar way, when the stream function S_k is prescribed at the nodal point ζ_k,
the approximate boundary value $\tilde{\Phi}(\zeta_k)$, given by the nodal equation for the potential
function, is equated with Φ_k. The approximate boundary value $\tilde{\Psi}(\zeta_k)$ is not forced to
be equal to the prescribed boundary value S_k; instead, the nodal value Ψ_k is equated
with S_k. Then, it follows that

$$
\begin{cases}
\tilde{\Phi}(\zeta_k) = \Phi_k \\
\Psi_k = S_k
\end{cases}
\tag{4.76}
$$

Substituting Eq. 4.76 into Eq. 4.62 gives

$$
\theta_k \Phi_k = S_k \ln\left|\frac{\zeta_{k+1} - \zeta_k}{\zeta_{k-1} - \zeta_k}\right| + \mathrm{Im}[G_k]
\tag{4.77}
$$

and Φ_k is left unknown.

In either of the boundary-condition cases, one of the nodal equations (Eq. 4.63
when P_k is given or Eq. 4.62 when S_k is given) is used. The result is a set of n_b
equations for n_b nodal points, which yields unique solutions for n_b unprescribed
boundary values.

4.4.2.1 Numerical Result

Formulation II applies the equivalence Eq. 4.74 along the boundary BCD and the
equivalence Eq. 4.76 along the boundary DA_1A_2B. Figure 4.23 shows the approx-
imate boundary values, $\tilde{\Phi}(\zeta_k)$ and $\tilde{\Psi}(\zeta_k)$, and computational errors, $\tilde{E}_\Phi(\zeta_k)$ and
$\tilde{E}_\Psi(\zeta_k)$, by the CVBEM with $n_b = 80$.

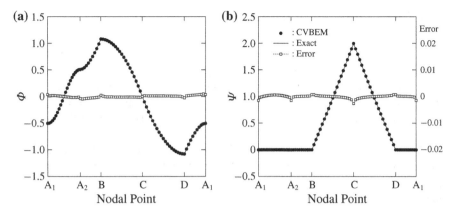

Fig. 4.23 Approximate boundary values, $\tilde{\Phi}(\zeta_k)$ and $\tilde{\Psi}(\zeta_k)$ (*black dots*), exact solutions, $\Phi(\zeta_k)$ and $\Psi(\zeta_k)$ (*solid lines*), and computational errors, $\tilde{E}_\Phi(\zeta_k)$ and $\tilde{E}_\Psi(\zeta_k)$ (*open squares*) for flow around a cylindrical corner using Formulation II with $n_b = 80$. **a** Velocity potential. **b** Stream function

Since Formulation II does not equate the approximate boundary values with the prescribed boundary values, the CVBEM yields errors at all the nodal points. However, the error profiles are relatively smooth; the severe oscillation observed with Formulation I (Fig. 4.21) is not present. Errors are more evenly distributed because the approximate boundary values are not forced to be equal to the prescribed boundary values.

The MAE characteristics with respect to n_b are shown in Fig. 4.22. The MAE of 2.06×10^{-2} for $n_b = 16$ is reduced to 3.71×10^{-3} by doubling n_b, which is further reduced to 5.24×10^{-4} for $n_b = 80$. The solution with 80 nodes is 39 times more accurate than the one with 16 nodes. Compared with Formulation I, Formulation II exhibits overall improvement in accuracy.

In Formulation II, the approximate boundary values, $\tilde{\Phi}(\zeta_k)$ or $\tilde{\Psi}(\zeta_k)$, are not forced to be equal to the prescribed boundary values, P_k or S_k. Hence, unlike the case in the known-variable equivalence (Formulation I), Formulation II can provide error estimations by using the difference between the approximate boundary values and prescribed boundary values:

$$\begin{cases} E_{\Phi k} = \tilde{\Phi}(\zeta_k) - P_k & \text{(on } \Gamma_P) \\ E_{\Psi k} = \tilde{\Psi}(\zeta_k) - S_k & \text{(on } \Gamma_S) \end{cases} \tag{4.78}$$

Since $P_k = \Phi(\zeta_k)$ and $S_k = \Psi(\zeta_k)$ (in this case), it follows that $E_{\Phi k} = \tilde{E}_\Phi(\zeta_k)$ and $E_{\Psi k} = \tilde{E}_\Psi(\zeta_k)$; the profiles of $E_{\Phi k}$ and $E_{\Psi k}$ are respectively the same as $\tilde{E}_\Phi(\zeta_k)$ and $\tilde{E}_\Psi(\zeta_k)$ shown in Fig. 4.23.

4.4.3 Dual-Variable Equivalence

Formulation I and Formulation II use either Eq. 4.62 or Eq. 4.63 at each of the nodal points, which is sufficient to yield a set of n_b equations for n_b unprescribed boundary values. The third approach uses Eqs. 4.62 and 4.63 simultaneously. In the dual-variable equivalence (Formulation III), the nodal equations for the prescribed and unprescribed variables are adopted and presumed to be equal to the prescribed boundary value and the nodal value of the same kind, respectively.

When the velocity potential P_k is prescribed at the nodal point ζ_k, the approximate boundary value $\tilde{\Phi}(\zeta_k)$, given by the nodal equation for the velocity potential, is equated with P_k, and the approximate boundary value $\tilde{\Psi}(\zeta_k)$, given by the nodal equation for the stream function, is equated with the nodal value Ψ_k:

$$\begin{cases} \tilde{\Phi}(\zeta_k) = P_k \\ \tilde{\Psi}(\zeta_k) = \Psi_k \end{cases} \tag{4.79}$$

Substituting Eq. 4.79 into Eqs. 4.62 and 4.63 gives

$$\begin{cases} 2\pi P_k = \Psi_k \ln \left| \dfrac{\zeta_{k+1} - \zeta_k}{\zeta_{k-1} - \zeta_k} \right| + (2\pi - \theta_k)\Phi_k + \text{Im}[G_k] \\ 0 = \Phi_k \ln \left| \dfrac{\zeta_{k+1} - \zeta_k}{\zeta_{k-1} - \zeta_k} \right| + \theta_k \Psi_k + \text{Re}[G_k] \end{cases} \tag{4.80}$$

In a similar way, when the stream function S_k is prescribed at the nodal point ζ_k, the approximate boundary values $\tilde{\Phi}(\zeta_k)$ and $\tilde{\Psi}(\zeta_k)$ are respectively equated with the nodal value Φ_k and the prescribed boundary value S_k:

$$\begin{cases} \tilde{\Phi}(\zeta_k) = \Phi_k \\ \tilde{\Psi}(\zeta_k) = S_k \end{cases} \tag{4.81}$$

Substituting Eq. 4.81 into Eqs. 4.62 and 4.63 gives

$$\begin{cases} 0 = \Psi_k \ln \left| \dfrac{\zeta_{k+1} - \zeta_k}{\zeta_{k-1} - \zeta_k} \right| - \theta_k \Phi_k + \text{Im}[G_k] \\ -2\pi S_k = \Phi_k \ln \left| \dfrac{\zeta_{k+1} - \zeta_k}{\zeta_{k-1} - \zeta_k} \right| - (2\pi - \theta_k)\Psi_k + \text{Re}[G_k] \end{cases} \tag{4.82}$$

In either of the boundary-condition cases, the nodal values Φ_k and Ψ_k are not forced to be equal to the prescribed boundary values and left unknown, resulting in $2n_b$ unknown boundary values in total. Unlike the two previous formulations, both boundary-value equations (Eqs. 4.62 and 4.63) are adopted. The result is a set of $2n_b$ equations for n_b nodal points, which yields unique solutions for $2n_b$ unknown boundary values.

These equations can be written in matrix form as

$$\mathbf{A}\Omega = \mathbf{Y} \tag{4.83}$$

$\mathbf{A}$ is the $2n_b \times 2n_b$ dense matrix that contains real and/or imaginary parts of H_j and I_j given by Eq. 4.66. Ω is the $2n_b$-dimensional vector that contains unknown nodal values:

$$\Omega = (\Phi_1, \Psi_1, \Phi_2, \Psi_2, \ldots, \Phi_{n_b}, \Psi_{n_b})^{\mathrm{T}} \tag{4.84}$$

$\mathbf{Y}$ is the $2n_b$-dimensional vector that contains $(2\pi P_k, 0)^{\mathrm{T}}$ if P_k is given or $(0, -2\pi S_k)^{\mathrm{T}}$ if S_k is given.

4.4.3.1 Numerical Result

Formulation III applies the equivalence Eq. 4.79 along the boundary BCD and the equivalence Eq. 4.81 along the boundary DA_1A_2B. Figure 4.24 shows the approximate boundary values, $\tilde{\Phi}(\zeta_k)$ and $\tilde{\Psi}(\zeta_k)$, and computational errors, $\tilde{E}_\Phi(\zeta_k)$ and $\tilde{E}_\Psi(\zeta_k)$, by the CVBEM with $n_b = 80$.

As is the case with Formulation I, Formulation III equates $\tilde{\Phi}(\zeta_k)$ with $P_k = \Phi(\zeta_k)$ along the boundary BCD, and the CVBEM yields no error in $\tilde{\Phi}(\zeta_k)$, as seen in Fig. 4.24a, while a small amount of error is observed in $\tilde{\Psi}(\zeta_k)$, as seen in Fig. 4.24b. The reverse is true along the boundary DA_1A_2B; no error in $\tilde{\Psi}(\zeta_k)$ and a small amount of error in $\tilde{\Phi}(\zeta_k)$.

The MAE characteristics with respect to n_b are shown in Fig. 4.22. The MAE of 1.28×10^{-2} for $n_b = 16$ is reduced to 2.56×10^{-3} by doubling n_b, which is further reduced to 3.19×10^{-4} for $n_b = 80$. The solution with 80 nodes is 40 times more

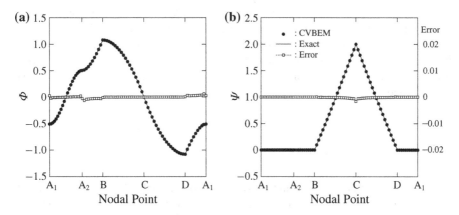

Fig. 4.24 Approximate boundary values, $\tilde{\Phi}(\zeta_k)$ and $\tilde{\Psi}(\zeta_k)$ (*black dots*), exact solutions, $\Phi(\zeta_k)$ and $\Psi(\zeta_k)$ (*solid lines*), and computational errors, $\tilde{E}_\Phi(\zeta_k)$ and $\tilde{E}_\Psi(\zeta_k)$ (*open squares*) for flow around a cylindrical corner using Formulation III with $n_b = 80$. **a** Velocity potential. **b** Stream function

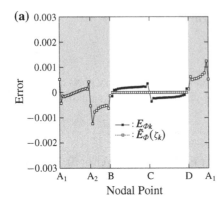 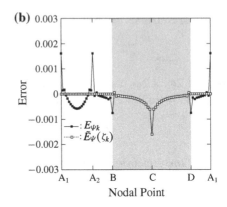

Fig. 4.25 Error estimations, $E_{\Phi k}$ and $E_{\Psi k}$ (*closed squares*), and computational errors, $\tilde{E}_{\Phi}(\zeta_k)$ and $\tilde{E}_{\Psi}(\zeta_k)$ (*open squares*) for flow around a cylindrical corner using Formulation III with $n_b = 80$. **a** Velocity potential. **b** Stream function

accurate than the one with 16 nodes. Formulation III yields further improvement in accuracy from Formulation II.

In Formulation III, the nodal values, Φ_k or Ψ_k, are not forced to be equal to the prescribed boundary values, P_k or S_k. Hence, Formulation III can provide error estimations by using the difference between the nodal values and prescribed boundary values:

$$\begin{cases} E_{\Phi k} = \Phi_k - P_k & \text{(on } \Gamma_P) \\ E_{\Psi k} = \Psi_k - S_k & \text{(on } \Gamma_S) \end{cases} \tag{4.85}$$

which give different values from $\tilde{E}_{\Phi}(\zeta_k)$ on Γ_P and $\tilde{E}_{\Psi}(\zeta_k)$ on Γ_S.

Figure 4.25 shows the profiles of $E_{\Phi k}$ along the boundary BCD and $E_{\Psi k}$ along the boundary DA_1A_2B. For comparison purposes, $\tilde{E}_{\Phi}(\zeta_k)$ and $\tilde{E}_{\Psi}(\zeta_k)$ are also shown. Note that the scale for errors is different (10 times magnified) from Fig. 4.24.

The values of $E_{\Phi k}$ and $E_{\Psi k}$, respectively, provide error estimations along the boundary BCD and DA_1A_2B, where $\tilde{E}_{\Phi}(\zeta_k)$ and $\tilde{E}_{\Psi}(\zeta_k)$ are of no use for error estimations. The magnitude of error is quite small and the accuracy of the CVBEM is confirmed.

4.5 Moduli of Functions

Error estimations discussed in the previous section are only along the boundary. It must be confirmed that such estimations are enough to ensure the accuracy of interior solutions. Here, the relevant theorems are introduced.

4.5.1 Mean Value Properties

Let $f(z)$ be analytic in a simply connected domain which contains the circle C_ρ of radius ρ with its center at z_0. From Cauchy's integral formula, $f(z_0)$ is given by

$$f(z_0) = \frac{1}{2\pi i} \oint_{C_\rho} \frac{f(z)}{z - z_0} dz \tag{4.86}$$

The circle C_ρ is parameterized by $z = z_0 + \rho e^{i\theta}$, and it follows that

$$\begin{cases} 1/(z - z_0) = e^{-i\theta}/\rho \\ dz = i\rho e^{i\theta} d\theta \end{cases} \tag{4.87}$$

Then, Eq. 4.86 becomes

$$\begin{aligned}
f(z_0) &= \frac{1}{2\pi i} \int_0^{2\pi} \frac{f(z_0 + \rho e^{i\theta}) e^{-i\theta}}{\rho} i\rho e^{i\theta} d\theta \\
&= \frac{1}{2\pi} \int_0^{2\pi} f(z_0 + \rho e^{i\theta}) d\theta
\end{aligned} \tag{4.88}$$

which states that the value of $f(z)$ at the center z_0 of the circle is the arithmetic mean of its values on the circle. This is known as Gauss's mean value theorem.

Theorem 4.6 (Gauss's mean value theorem) *If $f(z)$ is analytic in a simply connected domain D, the value of $f(z)$ at a point z_0 in D is equal to the mean value of $f(z)$ on any circle in D with its center at z_0.*

4.5.2 Maximum Modulus Theorem

Let $f(z)$ be analytic and nonconstant in a domain D and suppose that $|f(z)|$ has a maximum value M at a point z_0 in D. Since $f(z)$ is not constant, there must exist a circle C_ρ of radius ρ with its center at z_0 such that the interior of C_ρ lies in D and the inequality $|f(z)| < M$ holds at some point z_1 of C_ρ, as shown in Fig. 4.26.

By continuity of $f(z)$, for a finite arc $C_{\rho 1}$ of C_ρ which contains z_1, as shown in Fig. 4.26, the following inequality holds:

$$|f(z)| \leq M - m \tag{4.89}$$

Fig. 4.26 A domain D
containing a circular contour
C_ρ of radius ρ with its center
at z_0

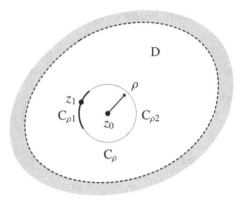

where $m > 0$. Let L_1 be the length of $C_{\rho 1}$ and L_2 be the length of the complementary
arc $C_{\rho 2}$. Since the length of C_ρ is $2\pi\rho$, $L_2 = 2\pi\rho - L_1$.

From Eq. 4.86 and the *ML* inequality, it follows that

$$M = |f(z_0)| \leq \frac{1}{2\pi} \left| \int_{C_{\rho 1}} \frac{f(z)}{z - z_0} dz \right| + \frac{1}{2\pi} \left| \int_{C_{\rho 2}} \frac{f(z)}{z - z_0} dz \right|$$

$$\leq \frac{1}{2\pi} \frac{M - m}{\rho} L_1 + \frac{1}{2\pi} \frac{M}{\rho} (2\pi\rho - L_1)$$

$$= M - \frac{L_1}{2\pi\rho} m < M \tag{4.90}$$

which is impossible. By virtue of this contradiction, it is concluded that the premise
is incorrect; that is, the assumption that $|f(z)|$ has a maximum value in D is invalid.

This leads to the following theorem, also known as the maximum modulus principle.

Theorem 4.7 (Maximum modulus theorem) *If $f(z)$ is analytic and nonconstant in
a domain D, the maximum value of $|f(z)|$ never occurs in D.*

If $f(z)$ is continuous throughout a closed bounded region R, $|f(z)|$ has a maximum
value somewhere in R. According to the maximum modulus principle, however,
the maximum value cannot occur in the interior of R. This leads to the following
important corollary.

Corollary 4.5 (Maximum modulus on a boundary) *If $f(z)$ is analytic and nonconstant in the interior of a closed region R and is continuous in R, the maximum value
of $|f(z)|$ in R occurs on the boundary of R.*

Furthermore, if $f(z) \neq 0$ anywhere in R, $1/f(z)$ is analytic there. From the
maximum modulus theorem, $1/|f(z)|$ cannot assume its maximum value in the

interior of R. Equivalently, $|f(z)|$ cannot assume its minimum value in the interior. Since $|f(z)|$ has a minimum, the minimum must be attained on the boundary of R. This leads to another corollary.

Corollary 4.6 (Minimum modulus on a boundary) *If $f(z)$ is analytic and nonconstant in the interior of a closed region R and is continuous and nonzero in R, the minimum value of $|f(z)|$ in R occurs on the boundary of R.*

Example 4.12 Let us confirm the maximum modulus theorem for $f(z) = e^z$, which is analytic, nonconstant, continuous, and nonzero in a bounded and closed region. Since $|e^z| = e^x$, the maximum of $|f(z)|$ occurs at a point in the region with maximal x and the minimum of $|f(z)|$ occurs at a point with minimal x. Both points are on the boundary of the region and not in the interior.

Example 4.13 Let us find the maximum and minimum moduli of $f(z) = z(z-1)$ in the circular region R given by $|z| \leq 1$. Since $f(z)$ is analytic and nonconstant in R, the maximum modulus theorem holds. The maximum value of $|f(z)|$ occurs on the boundary of R, which is found at $z = -1$ and the value is 2. Since $f(z) = 0$ at $z = 0$, which is in the interior of R, the minimum value of $|f(z)|$ can occur in the interior of R. Indeed, the minimum modulus is found at $z = 0$ and the value is 0.

With the CVBEM introduced in Sect. 4.3, the approximate solution $\tilde{\Omega}(z)$ is given by Eq. 4.45, which is analytic in the problem domain Π bounded by the boundary Γ. The exact solution $\Omega(z)$ is, of course, analytic, and thus, the error defined by

$$\tilde{E}(z) = \tilde{\Omega}(z) - \Omega(z) \tag{4.91}$$

is also analytic.

According to the maximum modulus principle, the maximum value of $|\tilde{E}(z)|$ must occur on the problem boundary Γ. Hence, error estimations along the boundary, such as Eq. 4.73, provide adequate information on the computational accuracy of the CVBEM throughout the problem region.

4.5.2.1 Numerical Result

The accuracy of the CVBEM must be assessed for interior solutions as well as for boundary solutions. With the CVBEM, the interior solutions $\tilde{\Omega}(z)$ for flow around a cylindrical corner, considered in Sect. 4.4, are computed at a number of evenly spaced points within the flow domain, and the errors $\tilde{E}(z)$ defined by Eq. 4.91 are evaluated.

Figure 4.27 shows the spatial distributions of $|\tilde{E}(z)|$ over the problem domain for Formulation I with $n_b = 16$ and $n_b = 80$. It should be noted that the grayscale is different between Fig. 4.27a, b. The accuracy of the interior solutions, as well as that of the boundary solutions, is improved as n_b increases. As can be deduced from the

(a)   **(b)**

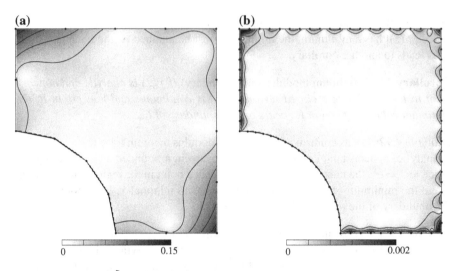

0   0.15   0   0.002

Fig. 4.27 Modulus of $\tilde{E}(z)$ within the flow domain for flow around a cylindrical corner. **a** Formulation I with $n_b = 16$. **b** Formulation I with $n_b = 80$

maximum modulus principle, the maximum error necessarily occurs on the boundary and larger errors are observed near the boundary. Away from the boundary, errors tend to decrease steadily.

Figure 4.28a, b show the spatial distributions of $|\tilde{E}(z)|$ for Formulations II and III, respectively, with $n_b = 80$. The errors diffuse over the problem domain, and the maximum value of $|\tilde{E}(z)|$ is found on the boundary in both cases. It is confirmed that the maximum modulus principle applies for the computational errors of the CVBEM, which implies that the acceptable errors on the boundary ensure the overall computational accuracy.

Figure 4.29 shows the contour plots of $\tilde{\Phi}(z)$ and $\tilde{\Psi}(z)$ computed by Formulation III with $n_b = 80$, which is found to be accurate enough for practical purposes. Similar plots can be obtained with the analytical solution Eq. 3.41, as shown in Fig. 3.23, which cannot be distinguished from Fig. 4.29 because of the high accuracy of the CVBEM.

In the subsequent part of this book, Formulation III is exclusively used because of its small magnitude of errors and the mild error distributions (no severe concentration of errors near the boundary).

• Solution to Task 5-1
To express the flow conditions for flow around right-angle impermeable faults without a cut-off wall (Fig. 4.1a), the Neumann boundary conditions are set as shown in Fig. 4.30. Along the no-flow boundary DAB, $S(\zeta) = 0$. Along the inlet boundary BC, $S(\zeta) = \Psi(y) = 2y$, while along the outlet boundary CD, $S(\zeta) = \Psi(x) = 2x$.

Since this is a Neumann boundary value problem, as mentioned in Sect. 4.1.2, a reference value of Φ needs to be set at a reference point on the boundary to ensure the uniqueness of the solution. In this problem, $\Phi = 0$ is set at the corner point A.

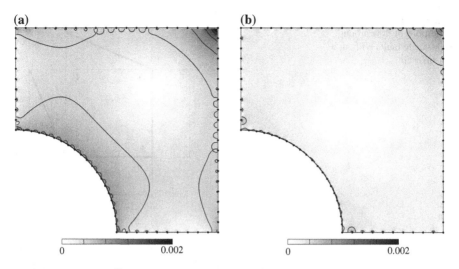

Fig. 4.28 Modulus of $\tilde{E}(z)$ within the flow domain for flow around a cylindrical corner. **a** Formulation II with $n_b = 80$. **b** Formulation III with $n_b = 80$

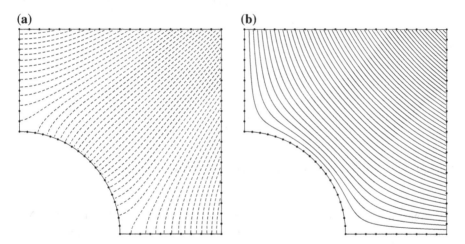

Fig. 4.29 Flow around a cylindrical corner using Formulation III with $n_b = 80$. **a** Equipotential lines. **b** Streamlines

In the CVBEM, the boundary ABCDA is modeled with 80 evenly spaced boundary elements of length 0.05. Figure 4.31 shows the contour plots of $\tilde{\Phi}(z)$ and $\tilde{\Psi}(z)$ computed by the CVBEM, which are the equipotential lines and streamlines, respectively.

The discharge $\Delta q / h$ between $(0, 0.5)$ and $(0.5, 0.5)$ can be evaluated as

$$\frac{\Delta q}{h} = \Psi(0.5 + 0.5\mathrm{i}) - \Psi(0.5\mathrm{i})$$

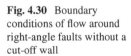

Fig. 4.30 Boundary
conditions of flow around
right-angle faults without a
cut-off wall

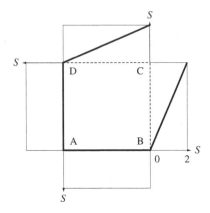

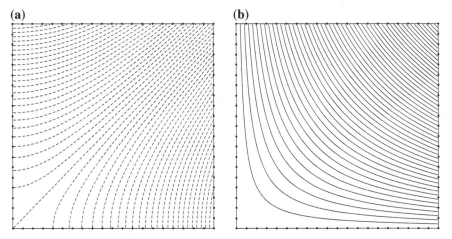

Fig. 4.31 Flow around right-angle faults. **a** Equipotential lines. **b** Streamlines

The stream function at $(0.5, 0.5)$ is obtained as $\tilde{\Psi}(0.5 + 0.5i) = 0.5$ by the CVBEM. Since $\Psi(0.5i) = S(\zeta) = 0$ as defined for the boundary condition, the discharge in the dry area is evaluated as $\Delta q / h = 0.5$.

The flow considered in Task 5-1 is the same as flow around a right-angle corner (Example 3.9) discussed in Sect. 3.2.2. The analytical solution is readily obtained by the conformal mapping as $\Omega = z^2$. The stream function at $(0.5, 0.5)$ is thus obtained analytically as 0.5, which confirms that the CVBEM result is accurate.

The spatial distribution of $|\tilde{E}(z)|$ over the problem domain is shown in Fig. 4.32. Note that the scale for errors is different (10 times magnified) from Fig. 4.28. The maximum value of $|\tilde{E}(z)|$ (2.94×10^{-4}) is found on the boundary, which is small

Fig. 4.32 Modulus of $\tilde{E}(z)$ within the flow domain for flow around a right-angle corner

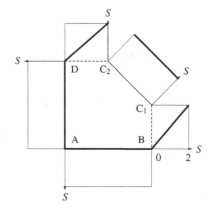

0   0.0002

Fig. 4.33 Boundary conditions of flow around right-angle faults with cut-off walls

enough, and the overall computational accuracy of the CVBEM is confirmed high enough for practical purposes.

• Solution to Task 5-2

To express the flow conditions for flow around right-angle impermeable faults with the cut-off walls (Fig. 4.1b), the Neumann boundary conditions are set as shown in Fig. 4.33. Along the no-flow boundary DAB, $S(\zeta) = 0$. Along the inlet boundary BC_1, $S(\zeta) = \Psi(y) = 4y$, while along the outlet boundary C_2D, $S(\zeta) = \Psi(x) = 4x$. The no-flow boundary C_1C_2 in between has a constant stream function $S(\zeta) = 2$. To ensure the uniqueness of the solution, $\Phi = 0$ is set at the corner point A.

In the CVBEM, the boundary C_2DABC_1 is modeled with 60 evenly spaced boundary elements of length 0.05, and the boundary C_1C_2 is modeled with 14 evenly spaced elements of length $\sqrt{2}/28$ ($\simeq 0.0505$). Figure 4.34 shows the contour plots of $\tilde{\Phi}(z)$ and $\tilde{\Psi}(z)$ computed by the CVBEM.

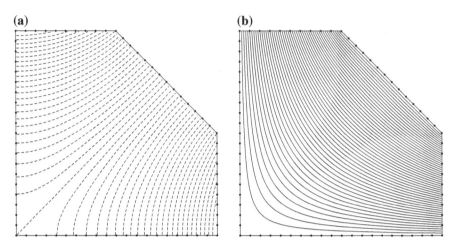

Fig. 4.34 Flow around right-angle faults with cut-off walls. **a** Equipotential lines. **b** Streamlines

Compared with the profiles for the case without a cut-off wall (Fig. 4.31), it is observed that the streamlines are shifted toward the boundary DAB and, consequently, the number of streamlines which pass through the dry area increases, implying an increased groundwater supply to the area.

With the CVBEM, the stream function at $(0.5, 0.5)$ for the case with the cut-off walls is obtained as $\tilde{\Psi}(0.5+0.5i) = 0.932$, and the discharge in the dry area becomes $\Delta q/h = 0.932$; around 86% increment of $\Delta q/h$ is expected by the installation of the cut-off walls.

Chapter 5
Singularity and Series

Motivating Problem 6: Artificial Fracture for Irrigation

Groundwater flows around right-angle impermeable faults with a flow velocity $q_u/A = 2$, as shown in Fig. 5.1. For the purpose of irrigation in the dry area, installation of an artificial fracture is planned. The fracture of length L_f, aperture w_f, and permeability k_f is to be created parallel to the x axis with its center at $(0.5, 0.4)$.

Task 6-1 For the fracture of $L_f = 0.5$, $w_f = 10^{-5}$, and $k_f = 10^7 k$, where k is the permeability of the flow medium, draw the groundwater flow profile and evaluate the discharge in the dry area $\Delta q / h$, represented by the amount flowing between $(0, 0.5)$ and $(0.5, 0.5)$.

Task 6-2 For the fracture of $w_f = 10^{-5}$ and $k_f = 10^7 k$, evaluate the relation between $\Delta q / h$ and the fracture length L_f.

Task 6-3 For the fracture of $L_f = 0.5$, evaluate the relation between $\Delta q / h$ and the fracture conductivity, defined by $k_f w_f$.

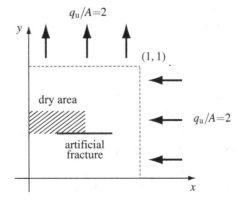

Fig. 5.1 Groundwater flowing through a dry area near right-angle impermeable faults and an artificial fracture

© Springer International Publishing Switzerland 2015
K. Sato, *Complex Analysis for Practical Engineering*,
DOI 10.1007/978-3-319-13063-7_5

• Solution Strategy to Motivating Problem 6

Flow with singular behavior around a fracture is considered in Chap. 3. For a simplified situation, such as uniform flow through an infinite conductivity fracture, the solution is sometimes available in a closed form. In practical situations, flow is arbitrary and fractures are not infinitely conductive, to which a closed-form solution may not always be found.

To achieve versatility in a solution method, a robust representation of flow with singular behavior is desired. In real-variable calculus, a closed-form function is often represented by power series expansions. The strategy is to develop the complex analog of the real power series to represent arbitrary flow through a finite conductivity fracture.

5.1 Sequences and Series

The convergence of complex sequences and series is first discussed as the basis for manipulating power series. The basic ideas of complex sequences and series are similar to those in real-variable calculus.

5.1.1 Convergence of Sequences

A sequence is a set of numbers, $z_1, z_2, z_3, \ldots$, also written as $\{z_n\}$, which is obtained by assigning a number z_n to each positive integer n. Each number in the sequence is called a term and z_n is called the nth term.

For any positive number ε, if there exists a positive number m such that

$$|z_n - z| < \varepsilon \quad \text{whenever} \quad m < n \tag{5.1}$$

a number z is called a limit of an infinite sequence $z_1, z_2, z_3, \ldots$. The value of m depends on the value of ε. Since ε can be arbitrarily small, the point z_n becomes arbitrarily close to z as n increases.

When a sequence has a limit z, the sequence is called convergent, written as

$$\lim_{n \to \infty} z_n = z \tag{5.2}$$

where the limit z is unique, that is, a sequence can converge to only one limit. If there exists no limit, the sequence is called divergent.

Example 5.1 Let us consider the sequence $\{0.9^n\}$. As n approaches ∞, 0.9^n tends to zero, and the sequence $\{0.9^n\}$ is convergent with the limit 0.

Example 5.2 Let us consider the sequence $\{i^n\} = i, -1, -i, 1, \ldots$. There exists no limit, and the sequence $\{i^n\}$ is divergent.

5.1.2 Convergence of Series

Let $z_1, z_2, z_3, \ldots$ be a given sequence. A new sequence $S_1, S_2, S_3, \ldots$ can be formed by

$$S_N = z_1 + z_2 + \cdots + z_N \tag{5.3}$$

where S_N is called the Nth partial sum of the following series:

$$\sum_{n=1}^{\infty} z_n = z_1 + z_2 + \cdots + z_n \cdots \tag{5.4}$$

When the sequence of partial sums $S_1, S_2, S_3, \ldots$ converges to S, a series is called convergent, written as

$$\sum_{n=1}^{\infty} z_n = S \tag{5.5}$$

where S is called the sum or value of the series. If a series does not converge, the series is called divergent.

Let us define the remainder U_N of the series after the term z_N as

$$U_N = z_{N+1} + z_{N+2} + \cdots = \sum_{n=N+1}^{\infty} z_n \tag{5.6}$$

If the series is convergent to S, it follows that

$$U_N = S - S_N \tag{5.7}$$

which implies that a series converges to S if and only if the reminder U_N tends to zero.

Furthermore, since $z_n = U_{n-1} - U_n$, it follows that

$$\lim_{n \to \infty} z_n = \lim_{n \to \infty} U_{n-1} - \lim_{n \to \infty} U_n \tag{5.8}$$

Therefore, if a series converges, then

$$\lim_{n \to \infty} z_n = 0 \tag{5.9}$$

must be satisfied. Note that Eq. 5.9 is a necessary condition for convergence; however, this is not sufficient.

Example 5.3 The harmonic series, a well-known divergent series, is defined by

$$\sum_{n=1}^{\infty} \frac{1}{n} = 1 + \frac{1}{2} + \frac{1}{3} + \cdots$$

Although Eq. 5.9 is satisfied, the series diverges.

Example 5.4 Let us consider the series $\sum_{n=1}^{\infty} z_n$, where $z_n = 0.9^{n-1}$. The Nth partial sum is

$$S_N = 1 + 0.9 + 0.9^2 + \cdots + 0.9^{N-1}$$

Multiplying both sides by 0.9 yields

$$0.9S_N = 0.9 + 0.9^2 + \cdots + 0.9^{N-1} + 0.9^N$$

By subtraction, it follows that

$$0.1S_N = 1 - 0.9^N$$

and S_N is obtained as

$$S_N = \frac{1 - 0.9^N}{0.1}$$

As N approaches ∞, 0.9^N tends to zero, and the sum of the series is found to be 10.

Example 5.5 Let us consider the series $\sum_{n=1}^{\infty} z_n$, where $z_n = 1.1^{n-1}$. With the same procedure as above, the Nth partial sum S_N is obtained as

$$S_N = \frac{-1 + 1.1^N}{0.1}$$

As N approaches ∞, 1.1^N tends to ∞, and the series is divergent.

5.1.3 Power Series

A power series in powers of $z - z_0$ is a series of the form

$$\sum_{n=0}^{\infty} \alpha_n (z - z_0)^n = \alpha_0 + \alpha_1 (z - z_0) + \alpha_2 (z - z_0)^2 + \cdots \qquad (5.10)$$

where the coefficients α_n and the center of the series z_0 are complex constants and z can be any point in a region containing z_0. Although power series are functions of a complex variable z, the ideas for series of constants discussed in the previous section are easily extended to Eq. 5.10.

As the series with constant terms converges or diverges, a series with a variable z converges for some values of z and diverges for others. Let us consider the smallest circle with its center at z_0 interior to which a given power series converges. Then, there exists a positive number R such that a series converges for $|z - z_0| < R$ and diverges for $|z - z_0| > R$. The circle given by

$$|z - z_0| = R \qquad (5.11)$$

is called the circle of convergence and its radius R is the radius of convergence. On the circle, a series may or may not converge.

Example 5.6 Let us consider the series $\sum_{n=1}^{\infty} z^{n-1}$, which replaces the constant terms in the series considered in Examples 5.4 and 5.5 by variable terms. If $|z| \geq 1$, from Eq. 5.9, the series is divergent. If $|z| < 1$, with the same procedure as before, the Nth partial sum $S_N(z)$ is obtained as

$$S_N(z) = \frac{1 - z^N}{1 - z} = \frac{1}{1 - z} - \frac{z^N}{1 - z}$$

As N approaches ∞, the last term tends to zero, and the convergent series is given by

$$\sum_{n=1}^{\infty} z^{n-1} = \frac{1}{1 - z} \qquad (5.12)$$

The radius of convergence is 1.

5.2 Taylor Series

In the following two sections, series representations of analytic functions are considered. The existence of such representations is guaranteed by the relevant theorems, which are the main results of the sections.

5.2.1 Taylor's Theorem

Every analytic function can be represented by a power series, known as a Taylor series.

Theorem 5.1 (Taylor's theorem) *If $f(z)$ is analytic throughout an open disk of radius R_0 with its center at z_0 given by $|z - z_0| < R_0$, then, at each point z in that disk, $f(z)$ has the series representation*

$$f(z) = \sum_{n=0}^{\infty} \frac{f^{(n)}(z_0)}{n!} (z - z_0)^n \qquad (5.13)$$

Fig. 5.2 Circular domain $|z - z_0| < R_0$ for a Taylor series

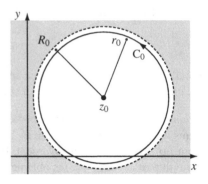

Proof Let C_0 be any positively oriented circle $|z - z_0| = r_0$, which is inside the disk $|z - z_0| < R_0$, as shown in Fig. 5.2. From Cauchy's integral formula

$$f(z) = \frac{1}{2\pi i} \oint_{C_0} \frac{f(w)}{w - z} dw \tag{5.14}$$

where z is interior to C_0.

An algebraic manipulation yields

$$\frac{1}{w - z} = \frac{1}{(w - z_0) - (z - z_0)} = \frac{1}{w - z_0} \frac{1}{1 - (z - z_0)/(w - z_0)}$$

$$= \frac{1}{w - z_0} \left[1 + \frac{z - z_0}{w - z_0} + \left(\frac{z - z_0}{w - z_0} \right)^2 + \cdots + \left(\frac{z - z_0}{w - z_0} \right)^{n-1} \right.$$

$$\left. + \left(\frac{z - z_0}{w - z_0} \right)^n \frac{1}{1 - (z - z_0)/(w - z_0)} \right] \tag{5.15}$$

where the following identity, for any complex number α other than unity, is used:

$$\frac{1}{1 - \alpha} = 1 + \alpha + \alpha^2 + \cdots + \alpha^{n-1} + \frac{\alpha^n}{1 - \alpha}$$

From Eqs. 5.14 and 5.15, it follows that

$$f(z) = \frac{1}{2\pi i} \oint_{C_0} \frac{f(w)}{w - z_0} dw + \frac{z - z_0}{2\pi i} \oint_{C_0} \frac{f(w)}{(w - z_0)^2} dw$$

$$+ \cdots + \frac{(z - z_0)^{n-1}}{2\pi i} \oint_{C_0} \frac{f(w)}{(w - z_0)^n} dw + U_n(z) \tag{5.16}$$

where

$$U_n(z) = \frac{1}{2\pi i} \oint_{C_0} \left(\frac{z - z_0}{w - z_0} \right)^n \frac{f(w)}{w - z} dw$$

From Cauchy's integral formula for derivatives

$$f^{(n)}(z_0) = \frac{n!}{2\pi i} \oint_{C_0} \frac{f(w)}{(w - z_0)^{n+1}} dw$$

Then, Eq. 5.16 is rewritten as

$$f(z) = f(z_0) + f'(z_0)(z - z_0) + \frac{f''(z_0)}{2!}(z - z_0)^2$$
$$+ \cdots + \frac{f^{(n-1)}(z_0)}{(n-1)!}(z - z_0)^{n-1} + U_n(z) \tag{5.17}$$

Since w is on C_0 and z is interior to C_0, it follows that

$$\left| \frac{z - z_0}{w - z_0} \right| = a < 1$$

and also

$$|w - z| = |(w - z_0) - (z - z_0)| \geq r_0 - |z - z_0|$$

Let M be the maximum value of $|f(w)|$ on C_0; then from the *ML* inequality

$$|U_n(z)| = \frac{1}{2\pi} \left| \oint_{C_0} \left(\frac{z - z_0}{w - z_0} \right)^n \frac{f(w)}{w - z} dw \right|$$
$$\leq \frac{1}{2\pi} \frac{a^n M}{r_0 - |z - z_0|} 2\pi r_0 = \frac{a^n M r_0}{r_0 - |z - z_0|}$$

which yields

$$\lim_{n \to \infty} U_n(z) = 0$$

Hence, Eq. 5.17 becomes

$$f(z) = f(z_0) + f'(z_0)(z - z_0) + \frac{f''(z_0)}{2!}(z - z_0)^2 + \cdots$$
$$= \sum_{n=0}^{\infty} \frac{f^{(n)}(z_0)}{n!}(z - z_0)^n$$

which is the desired result. □

Equation 5.13 is the expansion of $f(z)$ into a Taylor series about the point z_0. When $z_0 = 0$, the Taylor series reduces to a Maclaurin series:

$$f(z) = \sum_{n=0}^{\infty} \frac{f^{(n)}(0)}{n!} z^n \tag{5.18}$$

which is valid in the open disk $|z| < R_0$.

5.2.2 Special Taylor Series

Some important Taylor series expansions are reviewed. Note that the representations are the same as in real-variable calculus, since the coefficient formulae are the same.

5.2.2.1 Geometric Series

Let us consider a Maclaurin series for $f(z) = 1/(1 - z)$, which fails to be analytic at $z = 1$. The derivatives of $f(z)$ are $f^{(n)}(z) = n!/(1 - z)^{n+1}$ and $f^{(n)}(0) = n!$. From Eq. 5.18, it follows that

$$\frac{1}{1 - z} = \sum_{n=0}^{\infty} z^n \tag{5.19}$$

which is valid for $|z| < 1$. Note that Eq. 5.19 is consistent with Eq. 5.12.

5.2.2.2 Exponential Function

Let us consider a Maclaurin series for $f(z) = e^z$, which is analytic in the entire complex plane. The derivatives of $f(z)$ are $f^{(n)}(z) = e^z$ and $f^{(n)}(0) = 1$. From Eq. 5.18, it follows that

$$e^z = \sum_{n=0}^{\infty} \frac{z^n}{n!} \tag{5.20}$$

which is valid for $|z| < \infty$.

5.2.2.3 Trigonometric Functions

Let us consider a Maclaurin series for $f(z) = \sin z$, which is analytic in the entire complex plane. The derivatives of $f(z)$ are $f^{(2n+1)}(z) = (-1)^n \cos z$ and $f^{(2n)}(z) = (-1)^{n+1} \sin z$; thus, $f^{(2n+1)}(0) = (-1)^n$ and $f^{(2n)}(z) = 0$. From Eq. 5.18, it follows that

$$\sin z = \sum_{n=0}^{\infty} (-1)^n \frac{z^{2n+1}}{(2n+1)!} \tag{5.21}$$

which is valid for $|z| < \infty$.

In a similar way, for $f(z) = \cos z$, which is analytic in the entire complex plane, the derivatives of $f(z)$ are $f^{(2n+1)}(z) = (-1)^{n+1} \sin z$ and $f^{(2n)}(z) = (-1)^n \cos z$; thus, $f^{(2n+1)}(0) = 0$ and $f^{(2n)}(z) = (-1)^n$. From Eq. 5.18, it follows that

$$\cos z = \sum_{n=0}^{\infty} (-1)^n \frac{z^{2n}}{(2n)!} \tag{5.22}$$

which is valid for $|z| < \infty$.

5.2.2.4 Logarithmic Function

Let us consider a Maclaurin series for $f(z) = \ln(1 - z)$, which fails to be analytic at the branch point $z = 1$. The derivatives of $f(z)$ are $f^{(n)}(z) = -(n-1)!/(1-z)^n$ and $f^{(n)}(0) = -(n-1)!$ for $n \geq 1$. From Eq. 5.18, it follows that

$$\ln(1 - z) = -\sum_{n=1}^{\infty} \frac{z^n}{n} \tag{5.23}$$

which is valid for $|z| < 1$.

In a similar way, for $f(z) = \ln(1 + z)$, which fails to be analytic at the branch point $z = -1$. The derivatives of $f(z)$ are $f^{(n)}(z) = (-1)^{n-1}(n-1)!/(1+z)^n$ and $f^{(n)}(0) = (-1)^{n-1}(n-1)!$ for $n \geq 1$. From Eq. 5.18, it follows that

$$\ln(1 + z) = \sum_{n=1}^{\infty} (-1)^{n-1} \frac{z^n}{n} \tag{5.24}$$

which is valid for $|z| < 1$.

5.2.3 Uniqueness of Taylor Series Expansions

Mathematical manipulations in the coefficient formula sometimes become cumbersome. There are different ways to find Taylor series more easily than by the formula. Indeed, the geometric series in Example 5.6 is obtained without computing derivatives. The following theorem substantiates that the resultant series is the unique Taylor series, regardless of the method used.

Theorem 5.2 (Uniqueness of Taylor series expansions) *If a power series converges to $f(z)$ at all points in some open disk $|z - z_0| < R_0$, that is,*

$$f(z) = \sum_{n=0}^{\infty} \alpha_n (z - z_0)^n$$

then the series representation is the Taylor series expansion for $f(z)$ in powers of $z - z_0$.

Proof By the termwise differentiation, the derivative of the power series $f(z)$ of order m is obtained as

$$f^{(m)}(z) = \sum_{n=m}^{\infty} \frac{n!}{(n-m)!} \alpha_n (z - z_0)^{n-m}$$

For $z = z_0$, it follows that

$$f^{(m)}(z_0) = m! \alpha_m$$

and solving for α_m yields

$$\alpha_m = \frac{f^{(m)}(z_0)}{m!}$$

Then, the series $f(z)$ is given by

$$f(z) = \sum_{n=0}^{\infty} \alpha_n (z - z_0)^n = \sum_{n=0}^{\infty} \frac{f^{(n)}(z_0)}{n!} (z - z_0)^n$$

which is the Taylor series expansion of $f(z)$ and the theorem is proved. $\square$

Example 5.7 Let us find a Maclaurin series for $f(z) = 1/(1+z)$. Substituting $-z$ for z in Eq. 5.19 yields

$$\frac{1}{1+z} = \sum_{n=0}^{\infty} (-1)^n z^n \tag{5.25}$$

which is valid for $|z| < 1$.

Example 5.8 Let us find a Taylor series for $f(z) = 1/z$. Substituting $z - 1$ for z in Eq. 5.25 yields

$$\frac{1}{z} = \sum_{n=0}^{\infty} (-1)^n (z - 1)^n$$

which is valid for $|z - 1| < 1$.

Example 5.9 Let us find a Maclaurin series for $f(z) = 1/(1-z)^2$. From Eq. 5.19, $1/(1-z) = 1 + z + z^2 + \cdots$, for $|z| < 1$. Differentiation yields

$$\frac{1}{(1-z)^2} = 1 + 2z + 3z^2 + \cdots = \sum_{n=0}^{\infty}(n+1)z^n$$

which is valid for $|z| < 1$.

Example 5.10 Let us consider a Maclaurin series for $f(z) = \cos z$ by using the identity $\cos z = (e^{iz} + e^{-iz})/2$. From Eq. 5.20, it follows that

$$\cos z = \frac{1}{2}\sum_{m=0}^{\infty}\left(\frac{(iz)^m + (-iz)^m}{m!}\right) = 1 + 0 - \frac{z^2}{2!} + 0 + \frac{z^4}{4!} + \cdots = \sum_{n=0}^{\infty}(-1)^n\frac{z^{2n}}{(2n)!}$$

which is equivalent to Eq. 5.22.

By using the identity $\sin z = (e^{iz} - e^{-iz})/(2i)$ and Eq. 5.20, the series representation for $\sin z$ is obtained as Eq. 5.21.

Example 5.11 Let us consider a Maclaurin series for $f(z) = \ln(1-z)$ by using the geometric series. From Eq. 5.19, $1/(1-z) = 1 + z + z^2 + \cdots$, for $|z| < 1$. Integrating from 0 to z yields

$$-\ln(1-z) = z + \frac{z^2}{2} + \frac{z^3}{3} + \cdots = \sum_{n=1}^{\infty}\frac{z^n}{n}$$

which is equivalent to Eq. 5.23.

Example 5.12 Let us find a Maclaurin series for $f(z) = \ln[(1+z)/(1-z)]$. From Eqs. 5.23 and 5.24, it follows that

$$\ln\frac{1+z}{1-z} = 2\left(z + \frac{z^3}{3} + \frac{z^5}{5} + \cdots\right) = \sum_{n=0}^{\infty}\frac{2z^{2n+1}}{2n+1} \tag{5.26}$$

which is valid for $|z| < 1$.

5.3 Laurent Series

When $f(z)$ is not analytic at a point z_0, a Taylor series cannot be used to develop $f(z)$ in powers of $z - z_0$. However, it is often possible to find a series representation for $f(z)$ analytic in an annulus.

5.3.1 Laurent's Theorem

Laurent series, which consist of negative integer powers of $z - z_0$ as well as positive integer powers of $z - z_0$, generalize Taylor series.

Theorem 5.3 (Laurent's theorem) *If $f(z)$ is analytic throughout an annular domain $R_1 < |z - z_0| < R_2$, then, at each point z in that domain, $f(z)$ has the series representation*

$$f(z) = \sum_{n=-\infty}^{\infty} \alpha_n (z - z_0)^n \qquad (5.27)$$

where the power series converges to $f(z)$. The coefficients α_n are given by

$$\alpha_n = \frac{1}{2\pi i} \oint_C \frac{f(w)}{(w - z_0)^{n+1}} dw \qquad (5.28)$$

where n is an integer and C is any positively oriented simple closed contour around z_0 and lies in the domain.

Proof Let C_1 and C_2 be positively oriented circles $|z - z_0| = r_1$ and $|z - z_0| = r_2$, respectively, which form an annular region A, $r_1 \le |z - z_0| \le r_2$. The region A, contained in the domain $R_1 < |z - z_0| < R_2$, contains the contour C, as shown in Fig. 5.3. Since $f(z)$ is analytic in A, from Corollary 4.4, Cauchy's integral formula in the annular region A is given by

$$f(z) = \frac{1}{2\pi i} \oint_{C_2} \frac{f(w)}{w - z} dw - \frac{1}{2\pi i} \oint_{C_1} \frac{f(w)}{w - z} dw \qquad (5.29)$$

where z is contained in A.

Using Eq. 5.15, the first integral in Eq. 5.29 is rewritten as

Fig. 5.3 Annular domain $R_1 < |z - z_0| < R_2$ for a Laurent series

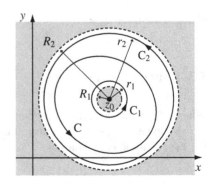

$$\frac{1}{2\pi i}\oint_{C_2}\frac{f(w)}{w-z}dw=\frac{1}{2\pi i}\oint_{C_2}\frac{f(w)}{w-z_0}dw+\frac{z-z_0}{2\pi i}\oint_{C_2}\frac{f(w)}{(w-z_0)^2}dw$$

$$+\cdots+\frac{(z-z_0)^{n-1}}{2\pi i}\oint_{C_2}\frac{f(w)}{(w-z_0)^n}dw+U_n(z)\qquad(5.30)$$

where

$$U_n(z)=\frac{1}{2\pi i}\oint_{C_2}\left(\frac{z-z_0}{w-z_0}\right)^n\frac{f(w)}{w-z}dw$$

As for the second integral in Eq. 5.29, it is noted that

$$-\frac{1}{w-z}=\frac{1}{z-w}$$

and thus, interchanging z and w in Eq. 5.15 gives

$$-\frac{1}{w-z}=\frac{1}{z-z_0}\left[1+\frac{w-z_0}{z-z_0}+\left(\frac{w-z_0}{z-z_0}\right)^2+\cdots+\left(\frac{w-z_0}{z-z_0}\right)^{n-1}\right.$$

$$+\left.\left(\frac{w-z_0}{z-z_0}\right)^n\frac{1}{1-(w-z_0)/(z-z_0)}\right]$$

Then, the second integral becomes

$$-\frac{1}{2\pi i}\oint_{C_1}\frac{f(w)}{w-z}dw=\frac{(z-z_0)^{-1}}{2\pi i}\oint_{C_1}f(w)dw+\frac{(z-z_0)^{-2}}{2\pi i}\oint_{C_1}\frac{f(w)}{(w-z_0)^{-1}}dw$$

$$+\cdots+\frac{(z-z_0)^{-n}}{2\pi i}\oint_{C_1}\frac{f(w)}{(w-z_0)^{-n+1}}dw+V_n(z)\qquad(5.31)$$

where

$$V_n(z)=\frac{1}{2\pi i}\oint_{C_1}\left(\frac{w-z_0}{z-z_0}\right)^n\frac{f(w)}{z-w}dw$$

From Eqs. 5.30 and 5.31, it follows that

$$f(z)=\sum_{j=0}^{n-1}\frac{(z-z_0)^j}{2\pi i}\oint_{C_2}\frac{f(w)}{(w-z_0)^{j+1}}dw$$

$$+\sum_{j=1}^{n}\frac{(z-z_0)^{-j}}{2\pi i}\oint_{C_1}\frac{f(w)}{(w-z_0)^{-j+1}}dw+U_n(z)+V_n(z)\qquad(5.32)$$

With the same method as used in the proof of Taylor's theorem, it can be shown that

$$\lim_{n \to \infty} U_n(z) = 0$$

As for $V_n(z)$, since w is on C_1 and z is contained in A, it follows that

$$\left| \frac{w - z_0}{z - z_0} \right| = a < 1$$

and also

$$|z - w| = |(z - z_0) - (w - z_0)| \geq |z - z_0| - r_1$$

Let M be the maximum value of $|f(w)|$ on C_1; then from the ML inequality

$$|V_n(z)| = \frac{1}{2\pi} \left| \oint_{C_1} \left(\frac{w - z_0}{z - z_0} \right)^n \frac{f(w)}{z - w} dw \right|$$

$$\leq \frac{1}{2\pi} \frac{a^n M}{|z - z_0| - r_1} 2\pi r_1 = \frac{a^n M r_1}{|z - z_0| - r_1}$$

which yields

$$\lim_{n \to \infty} V_n(z) = 0$$

Hence, Eq. 5.32 becomes

$$f(z) = \sum_{n=0}^{\infty} \frac{(z - z_0)^n}{2\pi i} \oint_{C_2} \frac{f(w)}{(w - z_0)^{n+1}} dw$$

$$+ \sum_{n=1}^{\infty} \frac{(z - z_0)^{-n}}{2\pi i} \oint_{C_1} \frac{f(w)}{(w - z_0)^{-n+1}} dw \qquad (5.33)$$

Since $f(z)$ is analytic in the annular region A, which contains the contour C, from Corollary 4.2, the contours C_1 and C_2 used in integrals can be deformed into the contour C, which completes the proof of Laurent's theorem. □

Equation 5.27 is the expansion of $f(z)$ into a Laurent series about the point z_0. The partial sum $\sum_{n=0}^{\infty} a_n(z - z_0)^n$ is called the analytic part of the Laurent series, while the remainder of the series consisting of inverse powers of $z - z_0$ is called the principal part.

Note that if the principal part is zero, the Laurent series reduces to a Taylor series. In Eq. 5.33, the first summation corresponds to the analytic part and the second corresponds to the principal part. When $f(z)$ is analytic throughout the disk $|z - z_0| < R_2$, the integrand of the principal part in Eq. 5.33, $f(w)(w - z_0)^{n-1}$, is also analytic, which results in the integral being zero. From Eq. 4.39, the analytic part is rewritten as

$$\sum_{n=0}^{\infty} \frac{(z-z_0)^n}{2\pi i} \oint_{C_2} \frac{f(w)}{(w-z_0)^{n+1}} dw = \sum_{n=0}^{\infty} \frac{f^{(n)}(z_0)}{n!} (z-z_0)^n \tag{5.34}$$

and the Laurent series reduces to a Taylor series about z_0.

Example 5.13 Let us find a Laurent series for $f(z) = 1/(1-z)$. Rearranging $f(z)$ and substituting $1/z$ for z in Eq. 5.19 yields

$$\frac{1}{1-z} = \frac{-1}{z} \frac{1}{1-1/z} = -\frac{1}{z} \sum_{n=0}^{\infty} \frac{1}{z^n} = -\sum_{n=0}^{\infty} \frac{1}{z^{n+1}} \tag{5.35}$$

which is valid for $|z| > 1$.

Example 5.14 Let us find a Laurent series for $f(z) = \ln[(z+1)/(z-1)]$. Rearranging $f(z)$ and substituting $1/z$ for z in Eq. 5.26 yields

$$\ln \frac{z+1}{z-1} = \ln \frac{1+1/z}{1-1/z} = 2 \left(\frac{1}{z} + \frac{1}{3z^3} + \frac{1}{5z^5} + \cdots \right)$$

$$= \sum_{n=0}^{\infty} \frac{2}{(2n+1)z^{2n+1}} \tag{5.36}$$

which is valid for $|z| > 1$.

Example 5.15 Let us consider the expansion of

$$f(z) = \frac{1}{(z-1)(z-2)} \tag{5.37}$$

into Maclaurin and Laurent series about $z_0 = 0$. In terms of partial fractions, $f(z)$ is rewritten as $-1/(z-1) + 1/(z-2)$, which has singularities at $z = 1$ and $z = 2$. Consequently, $f(z)$ is analytic in the open disk D; $|z| < 1$, in the annular domain A; $1 < |z| < 2$, and in the outer domain R; $2 < |z| < \infty$.

The series representation in D is a Maclaurin series. From Eq. 5.19, it follows that

$$f(z) = \frac{1}{1-z} - \frac{1}{2} \frac{1}{1-z/2} = \sum_{n=0}^{\infty} z^n - \sum_{n=0}^{\infty} \frac{z^n}{2^{n+1}}$$

$$= \sum_{n=0}^{\infty} (1 - 2^{-n-1}) z^n$$

which is valid for $|z| < 1$.

In the annular domain A, since $|1/z| < 1$ and $|z/2| < 1$ when $1 < |z| < 2$, from Eq. 5.19

$$f(z) = -\frac{1}{z}\frac{1}{1-1/z} - \frac{1}{2}\frac{1}{1-z/2} = -\sum_{n=0}^{\infty}\frac{1}{z^{n+1}} - \sum_{n=0}^{\infty}\frac{z^n}{2^{n+1}}$$

$$= -\sum_{n=0}^{\infty}\frac{z^n}{2^{n+1}} - \sum_{n=1}^{\infty}\frac{1}{z^n} \qquad (5.38)$$

which is the Laurent series valid for $1 < |z| < 2$.

The representation of $f(z)$ in R is also a Laurent series. Since $|1/z| < 1$ and $|2/z| < 1$ when $2 < |z| < \infty$, from Eq. 5.19

$$f(z) = -\frac{1}{z}\frac{1}{1-1/z} + \frac{1}{z}\frac{1}{1-2/z} = -\sum_{n=0}^{\infty}\frac{1}{z^{n+1}} + \sum_{n=0}^{\infty}\frac{2^n}{z^{n+1}}$$

$$= \sum_{n=1}^{\infty}\frac{2^{n-1}-1}{z^n} \qquad (5.39)$$

which is valid for $2 < |z| < \infty$.

Example 5.16 Let us consider the expansion of Eq. 5.37 into a Laurent series about singular points $z_0 = 1$ and 2. Let $w = z - 1$, then from Eq. 5.19

$$\frac{1}{(z-1)(z-2)} = \frac{1}{w(w-1)} = -\frac{1}{w}\frac{1}{1-w} = -\sum_{n=0}^{\infty}w^{n-1}$$

$$= -\sum_{n=0}^{\infty}(z-1)^n - \frac{1}{z-1} \qquad (5.40)$$

which is the Laurent series about $z_0 = 1$, valid for $0 < |z-1| < 1$.

As for the series representation about $z_0 = 2$, let $w = z - 2$, then from Eq. 5.19

$$\frac{1}{(z-1)(z-2)} = \frac{1}{(w+1)w} = \frac{1}{w}\frac{1}{1-(-w)} = \sum_{n=0}^{\infty}(-1)^n w^{n-1}$$

$$= \sum_{n=0}^{\infty}(-1)^{n+1}(z-2)^n + \frac{1}{z-2} \qquad (5.41)$$

which is valid for $0 < |z-2| < 1$.

5.3.2 Uniqueness of Laurent Series Expansions

In Examples 5.13 through 5.16, the coefficient formula, Eq. 5.28, is not used to find the coefficients α_n in the Laurent series expansions. The question at hand is, then, whether the coefficient formula gives completely different results from those obtained in the examples. This possibility is rejected by the following theorem.

Theorem 5.4 (Uniqueness of Laurent series expansions) *If a power series converges to $f(z)$ at all points in some annular domain $R_1 < |z - z_0| < R_2$, that is,*

$$f(z) = \sum_{n=-\infty}^{\infty} \alpha_n (z - z_0)^n$$

then the series representation is the Laurent series expansion for $f(z)$ in powers of $z - z_0$.

Proof Let us consider a function $g(z)$ as

$$g(z) = \frac{(z - z_0)^{-m-1}}{2\pi i}$$

where m is an integer. By the termwise integration, the integral of $g(z)f(z)$ over a circle C around the annulus with its center at z_0 is obtained as

$$\oint_C g(z)f(z)dz = \sum_{n=-\infty}^{\infty} \frac{\alpha_n}{2\pi i} \oint_C (z - z_0)^{n-m-1}dz \tag{5.42}$$

From Example 4.10, it follows that

$$\oint_C (z - z_0)^{n-m-1}dz = 2\pi i$$

only when $n = m$, otherwise the integral is zero. Thus, Eq. 5.42 reduces to

$$\alpha_n = \frac{1}{2\pi i} \oint_C \frac{f(z)}{(z - z_0)^{n+1}}dz$$

which is the coefficient in the Laurent series expansion, Eq. 5.28, and the theorem is proved.   □

It should be noted that the uniqueness is valid only for a given annulus. As seen in Example 5.15 (and the example below), there may exist different Laurent series in different concentric annuli.

Example 5.17 Let us find Laurent series about the origin for $f(z) = 1/(z^2 - z^3)$. Multiplying Eq. 5.19 by $1/z^2$ yields

$$\frac{1}{z^2 - z^3} = \frac{1}{z^2} \frac{1}{1-z} = \frac{1}{z^2} \sum_{n=0}^{\infty} z^n = \sum_{n=0}^{\infty} z^{n-2}$$

which is valid for $0 < |z| < 1$. On the other hand, multiplying Eq. 5.35 by $1/z^2$ yields

$$\frac{1}{z^2 - z^3} = \frac{1}{z^2} \frac{1}{1-z} = -\frac{1}{z^2} \sum_{n=0}^{\infty} \frac{1}{z^{n+1}} = -\sum_{n=0}^{\infty} \frac{1}{z^{n+3}}$$

which is valid for $|z| > 1$.

5.4 Singularities and Zeros

Laurent and Taylor series are of use to explore important properties of complex functions: singularities and zeros. In particular, poles, introduced in this section, play an essential role in practical applications, as revealed in the later sections.

5.4.1 Classification of Singularities

Let us revisit the singularity defined in Sect. 2.3.3. A point z_0 is called an isolated singular point of a function $f(z)$ if $f(z)$ is not analytic at z_0 and there is a punctured disk of radius δ with its center at z_0, $0 < |z - z_0| < \delta$, throughout which $f(z)$ is analytic. For instance, since $f(z) = 1/z$ is analytic everywhere except at $z_0 = 0$, the origin is an isolated singular point of $f(z) = 1/z$. As for $f(z) = \ln z$, however, the singularity at $z_0 = 0$ is not isolated since any neighborhood of z_0 contains points on the branch cut, at which $f(z)$ is not analytic.

In general, a function $f(z)$ with isolated singularities has a Laurent series since the punctured disk $0 < |z - z_0| < \delta$ is the same as the annular domain for expanding $f(z)$ in a Laurent series. Let $f(z)$ have an isolated singularity at z_0 with Laurent series

$$f(z) = \sum_{n=-\infty}^{\infty} \alpha_n (z - z_0)^n$$

$$= \sum_{n=0}^{\infty} \alpha_n (z - z_0)^n + \sum_{n=1}^{\infty} \frac{\alpha_{-n}}{(z - z_0)^n} \qquad (5.43)$$

By examination of Eq. 5.43, the isolated singularities of $f(z)$ can be classified into three types: removable, pole, and essential singularities. These singularities are mutually exclusive, that is, an isolated singularity is either a removable singularity, a pole, or an essential singularity.

5.4.1.1 Removable Singularities

If all of the coefficients α_{-n} in Eq. 5.43 are zero, the principal part of $f(z)$ at an isolated singular point z_0 vanishes and Eq. 5.43 takes the form

$$f(z) = \sum_{n=0}^{\infty} \alpha_n (z - z_0)^n \tag{5.44}$$

which is essentially a Taylor series. As there is no negative power term, the limit of $f(z)$ as z approaches z_0 exists and is equal to α_0. The isolated singularity at z_0 is called a removable singularity of $f(z)$, since it can be removed by defining $f(z_0) = \alpha_0$ and $f(z)$ becomes analytic at z_0.

Example 5.18 The function $f(z) = \sin z / z$ is undefined at $z_0 = 0$. The Laurent series of $f(z)$ is

$$f(z) = \frac{\sin z}{z} = \frac{1}{z}\left(z - \frac{z^3}{3!} + \frac{z^5}{5!} - \frac{z^7}{7!} + \cdots\right)$$
$$= 1 - \frac{z^2}{3!} + \frac{z^4}{5!} - \frac{z^6}{7!} + \cdots$$

which is valid for $0 < |z| < \infty$ and has no principal part. Hence, the singularity of $f(z)$ at $z_0 = 0$ can be removed by assigning a suitable value of $f(0) = 1$ at z_0.

Example 5.19 The function $f(z) = (e^z - 1)/z$ is undefined at $z_0 = 0$. The Laurent series of $f(z)$ is

$$f(z) = \frac{e^z - 1}{z} = \frac{1}{z}\left(1 + \frac{z}{1!} + \frac{z^2}{2!} + \frac{z^3}{3!} + \cdots - 1\right)$$
$$= 1 + \frac{z}{2!} + \frac{z^2}{3!} + \cdots$$

which is valid for $0 < |z| < \infty$ and has no principal part. Hence, the singularity of $f(z)$ at $z_0 = 0$ can be removed by assigning a suitable value of $f(0) = 1$ at z_0.

5.4.1.2 Poles

If the principal part of $f(z)$ at an isolated singular point z_0 contains at least one nonzero term but the number of such terms is finite, Eq. 5.43 takes the form

$$f(z) = \sum_{n=0}^{\infty} \alpha_n (z - z_0)^n + \frac{\alpha_{-1}}{z - z_0} + \frac{\alpha_{-2}}{(z - z_0)^2} + \cdots + \frac{\alpha_{-m}}{(z - z_0)^m} \quad (5.45)$$

where $\alpha_{-m} \neq 0$. This implies that there exists a positive integer m such that

$$\lim_{z \to z_0} (z - z_0)^m f(z) = \alpha_{-m} \neq 0 \quad (5.46)$$

and the isolated singularity at z_0 is called a pole of order m. When $m = 1$, it is called a simple pole. If $f(z)$ has a pole at $z = z_0$, $f(z)$ tends to ∞ as z approaches z_0.

Example 5.20 The function considered in Examples 5.15 and 5.16 (Eq. 5.37) has the Laurent series given by Eqs. 5.40 and 5.41. Hence, the singularities at $z_0 = 1$ and $z_0 = 2$ are simple poles.

Example 5.21 The function $f(z) = \sin z / z^2$ is undefined at $z_0 = 0$. The Laurent series of $f(z)$ is

$$f(z) = \frac{\sin z}{z^2} = \frac{1}{z^2} \left(z - \frac{z^3}{3!} + \frac{z^5}{5!} - \frac{z^7}{7!} + \cdots \right)$$
$$= \frac{1}{z} - \frac{z}{3!} + \frac{z^3}{5!} - \frac{z^5}{7!} + \cdots$$

which is valid for $0 < |z| < \infty$ and has the nonzero coefficient of z^{-1}. Hence, the singularity of $f(z)$ at $z_0 = 0$ is a simple pole.

Example 5.22 The function $f(z) = e^z / z^2$ is undefined at $z_0 = 0$. The Laurent series of $f(z)$ is

$$f(z) = \frac{e^z}{z^2} = \frac{1}{z^2} \left(1 + \frac{z}{1!} + \frac{z^2}{2!} + \frac{z^3}{3!} + \cdots \right)$$
$$= \frac{1}{z^2} + \frac{1}{z} + \frac{1}{2!} + \frac{z}{3!} + \frac{z^2}{4!} + \cdots$$

which is valid for $0 < |z| < \infty$ and has the nonzero coefficient of z^{-2}. Hence, the singularity of $f(z)$ at $z_0 = 0$ is a pole of order 2.

5.4.1.3 Essential Singularities

If the principal part of $f(z)$ at an isolated singular point z_0 contains an infinite number of nonzero terms as Eq. 5.43, the singularity at z_0 is called an essential singularity.

Example 5.23 The function $f(z) = \sin(1/z)$ is undefined at $z_0 = 0$. The Laurent series of $f(z)$ is

$$f(z) = \sin\frac{1}{z} = \frac{1}{z} - \frac{1}{3!z^3} + \frac{1}{5!z^5} - \frac{1}{7!z^7} + \cdots$$

which is valid for $0 < |z| < \infty$ and has an infinite number of negative power terms. Hence, the singularity of $f(z)$ at $z_0 = 0$ is an essential singularity.

Example 5.24 The function $f(z) = e^{1/z}$ is undefined at $z_0 = 0$. The Laurent series of $f(z)$ is

$$f(z) = e^{1/z} = 1 + \frac{1}{1!z} + \frac{1}{2!z^2} + \frac{1}{3!z^3} + \cdots$$

which is valid for $0 < |z| < \infty$ and has an infinite number of negative power terms. Hence, the singularity of $f(z)$ at $z_0 = 0$ is an essential singularity.

Example 5.25 The function considered in Example 5.15 (Eq. 5.37) has the Laurent series, Eqs. 5.38 and 5.39, with infinitely many negative power terms α_{-n}/z^n, which are only valid for $1 < |z| < 2$ and $2 < |z| < \infty$, respectively. Hence, it is incorrect to claim that the singularity at $z = 0$ is an essential singularity of $f(z)$. In classifying singularities, the Laurent series in the immediate neighborhood of a singularity must be used.

5.4.2 Zeros of Analytic Functions

A specific value z_0 for which $f(z_0) = 0$ is called a zero of a given function f. When $f(z)$ is analytic at a point z_0, it is analytic in some neighborhood of z_0, given by $|z - z_0| < R_0$, and from Taylor's theorem, it follows that

$$f(z) = \sum_{n=0}^{\infty} \alpha_n (z - z_0)^n \tag{5.47}$$

where $\alpha_n = f^{(n)}(z_0)/n!$. A function $f(z)$ is said to have a zero of order m at z_0, if and only if

$$\begin{cases} f^{(n)}(z_0) = 0 & \text{for} \quad n = 0, 1, 2, \ldots, m - 1 \\ f^{(m)}(z_0) \neq 0 \end{cases} \tag{5.48}$$

A zero with $m = 1$ is called a simple zero.

Example 5.26 Let us find the zeros of $f(z) = z^6 + z^4$ and evaluate their orders.

$$f(z) = z^4(z^2 + 1) = z^4(z - i)(z + i) \tag{5.49}$$

which implies that there are three zeros at $z_0 = 0, i,$ and $-i$. For $z_0 = 0$, it can be shown that $f'(z_0) = 6z_0^5 + 4z_0^3 = 0$, $f''(z_0) = 30z_0^4 + 12z_0^2 = 0$, $f^{(3)}(z_0) = 120z_0^3 + 24z_0 = 0$, and $f^{(4)}(z_0) = 360z_0^2 + 24 = 24 \neq 0$. Hence, $f(z)$ has a zero of order 4 at $z_0 = 0$. For $z_0 = i$, $f'(z_0) = 6z_0^5 + 4z_0^3 = 2i \neq 0$. Hence, $f(z)$ has a simple zero at $z_0 = i$. For $z_0 = -i$, $f'(z_0) = 6z_0^5 + 4z_0^3 = -2i \neq 0$. Hence, $f(z)$ has a simple zero at $z_0 = -i$.

Example 5.27 Let us find the zeros of $f(z) = z^2/e^z$ and evaluate their orders. There is a zero at $z_0 = 0$. For $z_0 = 0$, it can be shown that $f'(z_0) = 2z_0/e^{z_0} - z_0^2/e^{z_0} = 0$ and $f''(z_0) = 2/e^{z_0} - 4z_0/e^{z_0} + z_0^2/e^{z_0} = 2 \neq 0$. Hence, $f(z)$ has a zero of order 2 at $z_0 = 0$.

If $f(z)$ is analytic in a domain $|z - z_0| < R_0$ and has a zero of order m at z_0, from Taylor's theorem and the definition of zeros, $f(z)$ is given by

$$f(z) = \sum_{n=m}^{\infty} \alpha_n(z - z_0)^n = \sum_{n=0}^{\infty} \alpha_{n+m}(z - z_0)^{n+m}$$

$$= (z - z_0)^m \sum_{n=0}^{\infty} \alpha_{n+m}(z - z_0)^n \tag{5.50}$$

where $\alpha_n = f^{(n)}(z_0)/n!$ and $\alpha_m \neq 0$. Since convergent power series represent analytic functions, $f(z)$ can be written as

$$f(z) = (z - z_0)^m g(z) \tag{5.51}$$

where

$$g(z) = \sum_{n=0}^{\infty} \alpha_{n+m}(z - z_0)^n = \alpha_m + \sum_{n=1}^{\infty} \alpha_{n+m}(z - z_0)^n \tag{5.52}$$

is analytic at z_0 and $g(z_0) = \alpha_m \neq 0$.

Conversely, if $f(z)$ has the form given by Eq. 5.51, where $g(z)$ is analytic and nonzero at z_0, then $g(z)$ has the series representation as

$$g(z) = \sum_{n=0}^{\infty} \beta_n(z - z_0)^n \tag{5.53}$$

where $g(z_0) = \beta_0 \neq 0$. Multiplying both sides by $(z - z_0)^m$ yields

$$f(z) = g(z)(z - z_0)^m = \sum_{n=0}^{\infty} \beta_n (z - z_0)^{n+m}$$

$$= \sum_{n=m}^{\infty} \beta_{n-m} (z - z_0)^n \qquad (5.54)$$

Since $\beta_{m-m} = \beta_0 \neq 0$, $f(z)$ has a zero of order m at z_0.

These observations lead to the following theorem, which allows us to find zeros and their orders without evaluating derivatives.

Theorem 5.5 (Zero of order m) *When $f(z)$ is analytic in a domain $|z - z_0| < R_0$, $f(z)$ has a zero of order m at the point z_0 if and only if $f(z)$ can be expressed in the form*

$$f(z) = (z - z_0)^m g(z) \qquad (5.55)$$

where $g(z)$ is analytic at the point z_0 and $g(z_0) \neq 0$.

Example 5.28 Let us find the zeros of $f(z) = z^6 + z^4$ and evaluate their orders. From Eq. 5.49, it follows that

$$\begin{aligned} f(z) &= z^4(z^2 + 1) = z^4 g_1(z) \\ &= (z - i)(z^5 + iz^4) = (z - i)g_2(z) \\ &= (z + i)(z^5 - iz^4) = (z + i)g_3(z) \end{aligned}$$

For $z_0 = 0$, $g_1(z)$ is analytic at z_0 and $g_1(z_0) = 1 \neq 0$; thus, $z_0 = 0$ is a zero of order 4. For $z_0 = i$, $g_2(z)$ is analytic at z_0 and $g_2(z_0) = 2i \neq 0$; thus, $z_0 = i$ is a simple zero. For $z_0 = -i$, $g_3(z)$ is analytic at z_0 and $g_3(z_0) = -2i \neq 0$; thus, $z_0 = -i$ is a simple zero.

Example 5.29 Let us find the zeros of $f(z) = z^2/e^z$ and evaluate their orders. From Eq. 5.20, it follows that

$$f(z) = z^2 \left(1 - \frac{z}{1!} + \frac{z^2}{2!} - \frac{z^3}{3!} + \cdots \right) = z^2 g(z)$$

For $z_0 = 0$, $g(z)$ is analytic at z_0 and $g(z_0) = 1 \neq 0$; thus, $z_0 = 0$ is a zero of order 2.

5.4.3 Poles and Zeros

Poles are closely related to zeros. Suppose that $f(z)$ is analytic in an annular domain $0 < |z - z_0| < R_2$ and has a pole of order m at z_0. The Laurent series for $f(z)$ is given by

$$f(z) = \frac{1}{(z - z_0)^m} \sum_{n=0}^{\infty} \alpha_{n-m}(z - z_0)^n \tag{5.56}$$

where $\alpha_{-m} \neq 0$. The power series on the right-hand side represents an analytic function, and $f(z)$ can be written as

$$f(z) = \frac{h(z)}{(z - z_0)^m} \tag{5.57}$$

where

$$h(z) = \sum_{n=0}^{\infty} \alpha_{n-m}(z - z_0)^n = \alpha_{-m} + \sum_{n=1}^{\infty} \alpha_{n-m}(z - z_0)^n \tag{5.58}$$

is analytic at z_0 and $h(z_0) = \alpha_{-m} \neq 0$.

Conversely, suppose that $f(z)$ has the form given by Eq. 5.57, where $h(z)$ is analytic and nonzero at z_0, then $h(z)$ has the series representation as

$$h(z) = \sum_{n=0}^{\infty} \beta_n (z - z_0)^n \tag{5.59}$$

where $h(z_0) = \beta_0 \neq 0$. Dividing both sides by $(z - z_0)^m$ yields

$$f(z) = \frac{h(z)}{(z - z_0)^m} = \sum_{n=0}^{\infty} \beta_n (z - z_0)^{n-m}$$

$$= \sum_{n=-m}^{\infty} \beta_{n+m}(z - z_0)^n \tag{5.60}$$

Since $\beta_{-m+m} = \beta_0 \neq 0$, $f(z)$ has a pole of order m at z_0.

These observations lead to the following theorem and corollary, which allow us to find poles and their orders by using the zeros of relevant functions.

Theorem 5.6 (Pole of order m) *When $f(z)$ is analytic in a domain $0 < |z - z_0| < R_2$, $f(z)$ has a pole of order m at the point z_0 if and only if $f(z)$ can be expressed in the form*

$$f(z) = \frac{h(z)}{(z - z_0)^m} \tag{5.61}$$

where $h(z)$ is analytic at the point z_0 and $h(z_0) \neq 0$.

Corollary 5.1 (Poles from zeros) *If $f(z)$ is analytic and has a zero of order m at the point z_0, $g(z) = 1/f(z)$ has a pole of order m at z_0.*

Example 5.30 Let us consider $f(z) = (z - 1)(z - 2)$, which has simple zeros at 1 and 2. From Corollary 5.1, $g(z) = 1/(z - 1)(z - 2)$ has simple poles at 1 and 2, which is consistent with the result of Example 5.20.

Example 5.31 Let us consider $f(z) = z^2/e^z$. From Example 5.27, $f(z)$ has a zero of order 2 at 0. From Corollary 5.1, $g(z) = e^z/z^2$ has a pole of order 2 at 0, which is consistent with the result of Example 5.22.

Let us consider two functions $p(z)$ and $q(z)$, which are analytic at a point z_0 and $p(z_0) \neq 0$. If $q(z)$ has a zero of order m at z_0, from Eq. 5.55

$$q(z) = (z - z_0)^m g(z) \tag{5.62}$$

where $g(z)$ is analytic at z_0 and $g(z_0) \neq 0$. Then the quotient $p(z)/q(z)$ becomes

$$\frac{p(z)}{q(z)} = \frac{p(z)/g(z)}{(z - z_0)^m} \tag{5.63}$$

Since $p(z)$ and $g(z)$ are analytic and nonzero at z_0, the same is true for the quotient $p(z)/g(z)$. Hence, Eq. 5.63 implies that the quotient $p(z)/q(z)$ has a pole of order m at z_0.

Conversely, if the quotient $p(z)/q(z)$ has a pole of order m at z_0, from Eq. 5.61

$$\frac{p(z)}{q(z)} = \frac{h(z)}{(z - z_0)^m} \tag{5.64}$$

where $h(z)$ is analytic at z_0 and $h(z_0) \neq 0$. Then $q(z)$ becomes

$$q(z) = (z - z_0)^m \frac{p(z)}{h(z)} \tag{5.65}$$

Since $p(z)$ and $h(z)$ are analytic and nonzero at z_0, the same is true for the quotient $p(z)/h(z)$. Hence, Eq. 5.65 reveals that $q(z)$ has a zero of order m at z_0.

These observations can be summarized by the following corollary.

Corollary 5.2 (Poles of a quotient of functions) *When $p(z)$ and $q(z)$ are analytic at the point z_0 and $p(z_0) \neq 0$, the quotient $p(z)/q(z)$ has a pole of order m at z_0 if and only if $q(z)$ has a zero of order m at z_0.*

Example 5.32 Let us consider $f(z) = \sin z/z^3$. For the case of

$$\begin{cases} p(z) = \sin z \\ q(z) = z^3 \end{cases}$$

where $p(z)$ and $q(z)$ are analytic at $z = 0$ but $p(0) = 0$, Corollary 5.2 cannot be applied. On the other hand, for the case of

$$\begin{cases} p(z) = \sin z / z \\ q(z) = z^2 \end{cases}$$

where $p(z)$ and $q(z)$ are analytic at $z = 0$ and $p(0) = \sin 0/0 = 1 \neq 0$ as discussed in Example 5.18, Corollary 5.2 can be applied. The function $q(z) = z^2$ has a zero of order 2 at $z = 0$, and it follows that $f(z)$ has a pole of order 2 at 0. For the case of

$$\begin{cases} p(z) = \sin z / z^2 \\ q(z) = z \end{cases}$$

where $p(z)$ is not analytic at $z = 0$, Corollary 5.2 cannot be applied.

5.5 Residue Theory

Cauchy's integral theorem, discussed in Sect. 4.2.4, relies on the analyticity of the integrand $f(z)$ within and on a closed contour C and states that the value of the contour integral of $f(z)$ is zero. Conversely, when $f(z)$ has singularities interior to C, the integral may yield nonzero value. This property is of practical use in evaluating complex and real integrals.

5.5.1 Residues

If $f(z)$ has a nonremovable isolated singularity at a point $z = z_0$, there exists a positive number R_2 such that $f(z)$ is analytic at each point in a domain $0 < |z - z_0| < R_2$. Then, $f(z)$ has a Laurent series

$$f(z) = \sum_{n=0}^{\infty} \alpha_n (z - z_0)^n + \frac{\alpha_{-1}}{z - z_0} + \frac{\alpha_{-2}}{(z - z_0)^2} + \cdots \qquad (5.66)$$

which is valid for $0 < |z - z_0| < R_2$. The coefficients α_{-n} are given by

$$\alpha_{-n} = \frac{1}{2\pi i} \oint_C \frac{f(w)}{(w - z_0)^{-n+1}} dw \qquad (5.67)$$

where $n = 1, 2, \ldots$ and C is any positively oriented simple closed contour around z_0 and lying in the domain $0 < |z - z_0| < R_2$.

With $n = 1$, the coefficient of $1/(z - z_0)$ in Eq. 5.66 is obtained as

$$\alpha_{-1} = \frac{1}{2\pi i} \oint_C f(w)dw \tag{5.68}$$

which is called the residue of $f(z)$ at the isolated singular point z_0. Let us denote the residue of $f(z)$ at z_0 by

$$\operatorname*{Res}_{z=z_0} f(z) = \frac{1}{2\pi i} \oint_C f(z)dz \tag{5.69}$$

Equation 5.69 provides a formula for evaluating certain integrals around simple closed contours C:

$$\oint_C f(z)dz = 2\pi i \operatorname*{Res}_{z=z_0} f(z) \tag{5.70}$$

Example 5.33 Let us evaluate the integral of $f(z) = e^{1/z}$ around the positively oriented unit circle C, $|z| = 1$. From Example 5.24, the Laurent series is given by

$$f(z) = e^{1/z} = 1 + \frac{1}{z} + \frac{1}{2!z^2} + \frac{1}{3!z^3} + \cdots$$

which is valid for $0 < |z| < \infty$. This series shows that $f(z)$ has a nonremovable isolated (essential) singularity at $z = 0$, which is interior to C, and the residue is 1. From Eq. 5.70, it follows that

$$\oint_C e^{1/z}dz = 2\pi i$$

Example 5.34 Let us evaluate the integral of $f(z) = e^{1/z^2}$ around the positively oriented unit circle C, $|z| = 1$. The Laurent series is given by

$$f(z) = e^{1/z^2} = 1 + \frac{1}{z^2} + \frac{1}{2!z^4} + \frac{1}{3!z^6} + \cdots$$

which is valid for $0 < |z| < \infty$. This series shows that $f(z)$ has a nonremovable isolated (essential) singularity at $z = 0$, which is interior to C, and the residue is 0. From Eq. 5.70, it follows that

$$\oint_C e^{1/z^2}dz = 0$$

This example shows that the analyticity of a function at all points interior to and on a simple closed contour C is a sufficient condition for the value of the integral around C to be 0 (Cauchy's integral theorem) but not a necessary condition.

Example 5.35 Let us evaluate the integral of $f(z) = e^{2z}/(z-1)^3$ around the positively oriented circle C, $|z| = 2$. Let $w = z - 1$, then $z = 1 + w$ and

$$\frac{e^{2z}}{(z-1)^3} = \frac{e^{2+2w}}{w^3} = \frac{e^2}{w^3}\left[1 + \frac{2w}{1!} + \frac{(2w)^2}{2!} + \frac{(2w)^3}{3!} + \frac{(2w)^4}{4!} + \cdots\right]$$

Then, the Laurent series is given by

$$f(z) = \frac{e^{2z}}{(z-1)^3} = \frac{e^2}{(z-1)^3} + \frac{2e^2}{(z-1)^2} + \frac{2e^2}{z-1} + \frac{4e^2}{3} + \frac{2e^2}{3}(z-1) + \cdots$$

which is valid for $0 < |z - 1| < \infty$. This series shows that $f(z)$ has a nonremovable isolated singularity (a pole of order 3) at $z = 1$, which is interior to C, and the residue is $2e^2$. From Eq. 5.70, it follows that

$$\oint_C \frac{e^{2z}}{(z-1)^3}dz = 4e^2\pi i$$

Note that if the radius of contour C is less than 1, then the isolated singularity lies exterior to C and the integral becomes zero.

5.5.2 Residues at Poles

A Laurent series expansion usually requires tedious calculations. When the isolated singularities are poles, however, there are some formulae useful for evaluating residues at poles.

5.5.2.1 Functions with a Simple Pole

If two functions $p(z)$ and $q(z)$ are analytic at a point z_0 and $f(z) = p(z)/q(z)$ has a simple pole at z_0, then it follows $p(z_0) \neq 0$ and $q(z_0) = 0$. Since $q(z)$ has a simple zero at z_0, $q'(z_0) \neq 0$, and from Eq. 5.55, $q(z)$ is expressed as

$$q(z) = (z - z_0)g(z) \tag{5.71}$$

where $g(z)$ is analytic at z_0 and $g(z_0) \neq 0$. Then, $f(z)$ becomes

$$f(z) = \frac{p(z)}{q(z)} = \frac{p(z)/g(z)}{z - z_0} \tag{5.72}$$

and the residue of $f(z)$ at z_0 is $p(z_0)/g(z_0)$. The Taylor series of $q(z)$ at z_0 is

$$q(z) = (z - z_0)q'(z_0) + \frac{(z - z_0)^2}{2!}q''(z_0) + \cdots \tag{5.73}$$

Substituting Eq. 5.73 into Eq. 5.71 and setting $z = z_0$ yields $g(z_0) = q'(z_0)$. Finally, the residue of $f(z) = p(z)/q(z)$ at a simple pole z_0 is given by

$$\operatorname*{Res}_{z=z_0} f(z) = \frac{p(z_0)}{q'(z_0)} \tag{5.74}$$

Example 5.36 Let us evaluate the residues of $f(z) = 1/(e^z - 1) = p(z)/q(z)$. Since $p(z) = 1$ and $q(z) = e^z - 1$ are analytic and $q(2n\pi i) = 0$ and $q'(2n\pi i) = e^{2n\pi i} \neq 0$, $f(z)$ has simple poles at $z = 2n\pi i$ for any integer n. From Eq. 5.74, the residues of $f(z)$ at $z = 2n\pi i$ are

$$\operatorname*{Res}_{z=2n\pi i} f(z) = \frac{1}{e^{2n\pi i}} = 1$$

5.5.2.2 Functions Expressing a Simple Pole

Suppose that $f(z)$ can be expressed in the form

$$f(z) = \frac{h(z)}{z - z_0} \tag{5.75}$$

where $h(z)$ is analytic at z_0 and $h(z_0) \neq 0$. Since $h(z)$ is analytic at z_0, it has the Taylor series expansion

$$h(z) = h(z_0) + \frac{h'(z_0)}{1!}(z - z_0) + \frac{h''(z_0)}{2!}(z - z_0)^2 + \cdots \tag{5.76}$$

which is valid for $|z - z_0| < R_2$, and it follows that

$$f(z) = \frac{h(z_0)}{z - z_0} + \frac{h'(z_0)}{1!} + \frac{h''(z_0)}{2!}(z - z_0) + \cdots \tag{5.77}$$

which is valid for $0 < |z - z_0| < R_2$. Equation 5.77 is the Laurent series representation for $f(z)$ and implies that $f(z)$ has a simple pole at z_0. The residue is readily obtained as

$$\operatorname*{Res}_{z=z_0} f(z) = h(z_0) \tag{5.78}$$

The same result is obtained through Eq. 5.74.

Example 5.37 Let us evaluate the residues of $f(z) = 1/(z-1)(z-2)$. Since $f(z)$ has the form of Eq. 5.75

$$f(z) = \frac{1}{(z-1)(z-2)} = \frac{1/(z-2)}{z-1} = \frac{h(z)}{z-1}$$

where $h(z) = 1/(z-2)$ is analytic at $z = 1$ and $h(1) \neq 0$, from Eq. 5.78, it follows that

$$\underset{z=1}{\text{Res}} f(z) = h(1) = -1$$

Similarly, $f(z)$ can be rewritten as

$$f(z) = \frac{1}{(z-1)(z-2)} = \frac{1/(z-1)}{z-2} = \frac{h(z)}{z-2}$$

where $h(z) = 1/(z-1)$ is analytic at $z = 2$ and $h(2) \neq 0$. From Eq. 5.78, it follows that

$$\underset{z=2}{\text{Res}} f(z) = h(2) = 1$$

These results are consistent with the residues obtained from the Laurent series in Example 5.16.

5.5.2.3 Functions Expressing a Pole of Order m

The same argument above can be applied to poles of higher order m. Suppose that $f(z)$ can be expressed in the form

$$f(z) = \frac{h(z)}{(z-z_0)^m} \tag{5.79}$$

where $h(z)$ is analytic at z_0 and $h(z_0) \neq 0$. Since $h(z)$ is analytic at z_0, it can be expressed as Eq. 5.76 and $f(z)$ is given by

$$f(z) = \frac{h(z_0)}{(z-z_0)^m} + \frac{h'(z_0)}{1!} \frac{1}{(z-z_0)^{m-1}} + \frac{h''(z_0)}{2!} \frac{1}{(z-z_0)^{m-2}}$$
$$+ \cdots + \frac{h^{(m-1)}(z_0)}{(m-1)!} \frac{1}{z-z_0} + \cdots \tag{5.80}$$

which is valid for $0 < |z - z_0| < R_2$. Equation 5.80 is the Laurent series representation for $f(z)$ and implies that $f(z)$ has a pole of order m at z_0. The residue is obtained as

$$\underset{z=z_0}{\text{Res}} f(z) = \frac{h^{(m-1)}(z_0)}{(m-1)!} \tag{5.81}$$

which reduces to Eq. 5.78 when $m = 1$.

Example 5.38 Let us evaluate the residues of $f(z) = e^{2z}/(z-1)^3$. Since $f(z)$ has the form of Eq. 5.79

$$f(z) = \frac{e^{2z}}{(z-1)^3} = \frac{h(z)}{(z-1)^3}$$

where $h(z) = e^{2z}$ is analytic at $z = 1$ and $h(1) \neq 0$, from Eq. 5.81, it follows that

$$\operatorname*{Res}_{z=1} f(z) = \frac{h''(1)}{2!} = \frac{4e^2}{2} = 2e^2$$

which is consistent with the result of Example 5.35.

5.5.3 Residue Theorem

Residue integration for the case with a single singularity can be extended to the case with multiple isolated singularities interior to the contour.

Theorem 5.7 (Residue theorem) *If $f(z)$ is analytic within and on a simple closed contour C except for a finite number of singular points z_j ($j = 1, 2, \ldots, n$) interior to C, then*

$$\oint_C f(z)\mathrm{d}z = 2\pi\mathrm{i} \sum_{j=1}^{n} \operatorname*{Res}_{z=z_j} f(z) \tag{5.82}$$

where C is in the counterclockwise direction.

Proof Let each of the singular points z_j ($j = 1, 2, \ldots, n$) be enclosed in a positively oriented circle C_j which is interior to C and is so small that those n circles and C are all separated, as shown in Fig. 5.4 for the case $n = 3$.

Fig. 5.4 Simple closed contour C containing circular contours C_j, respectively, enclosing isolated singular points z_j

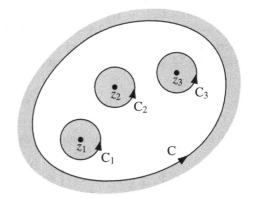

The function $f(z)$ is analytic in the multiply-connected domain bounded by C and C_j and on the entire boundary of the domain. From Theorem 4.3, it follows that

$$\oint_C f(z)dz - \sum_{j=1}^{n} \oint_{C_j} f(z)dz = 0 \tag{5.83}$$

where the second term on the left-hand side is given by Eq. 5.70 as

$$\sum_{j=1}^{n} \oint_{C_j} f(z)dz = 2\pi i \sum_{j=1}^{n} \operatorname*{Res}_{z=z_j} f(z) \tag{5.84}$$

Equations 5.83 and 5.84 reduce to Eq. 5.82 and the proof is complete.   □

Example 5.39 Let us evaluate the integral of $f(z) = 1/(z-1)(z-2)$ around the positively oriented circle C, $|z| = 3$. The integrand has two singularities at $z = 1$ and 2, both of which are interior to C. From Example 5.37, the residues are -1 at $z = 1$ and 1 at $z = 2$. Hence, from the residue theorem

$$\oint_C \frac{1}{(z-1)(z-2)}dz = 2\pi i(-1+1) = 0$$

Note that if the radius of contour C is between 1 and 2, the singular point $z = 2$ lies exterior to C and the integral becomes

$$\oint_C \frac{1}{(z-1)(z-2)}dz = 2\pi i(-1) = -2\pi i$$

and if the radius is less than 1, there is no singularity interior to C and the integral becomes zero.

Example 5.40 Let us evaluate the integral of $f(z) = 1/(z^4+4)$ around the positively oriented circle C, $|z-1| = 2$. The denominator of $f(z)$ has four simple zeros at $z = 1 \pm i$ and $z = -1 \pm i$. The first two are interior to C and the remaining two are exterior to C. Hence, the integrand $f(z)$ has two simple poles at $z = 1 \pm i$ within C. From Eq. 5.74, the residues of $f(z)$ at $z = 1 \pm i$ are

$$\operatorname*{Res}_{z=1\pm i} f(z) = \frac{1}{4(1 \pm i)^3}$$

From the residue theorem, the integral is evaluated as

$$\oint_C \frac{1}{z^4+4}dz = 2\pi i \left[\frac{1}{4(1+i)^3} + \frac{1}{4(1-i)^3} \right] = -\frac{\pi}{4}i$$

Note that if the radius of contour C is less than 1, there is no singularity interior to C and the integral becomes zero.

5.5.4 Residue Integration of Real Integrals

A practical application of the residue theory is the evaluation of certain classes of definite and improper real integrals.

5.5.4.1 Definite Integrals Involving Trigonometric Functions

Let us consider integrals of $f(\theta)$, which is a function of $\cos\theta$ and $\sin\theta$, over the interval $0 \leq \theta \leq 2\pi$. By setting $e^{i\theta} = z$, it follows that

$$
\begin{cases}
\cos\theta = \dfrac{1}{2}(e^{i\theta} + e^{-i\theta}) = \dfrac{1}{2}\left(z + \dfrac{1}{z}\right) \\[3mm]
\sin\theta = \dfrac{1}{2i}(e^{i\theta} - e^{-i\theta}) = \dfrac{1}{2i}\left(z - \dfrac{1}{z}\right)
\end{cases}
\tag{5.85}
$$

Since $dz/d\theta = ie^{i\theta} = iz$, the integral of $f(\theta)$ takes the form

$$
\int_0^{2\pi} f(\theta)d\theta = \oint_C F(z)dz
\tag{5.86}
$$

where C is the positively oriented unit circle, $|z| = 1$, and $F(z) = f(\theta)/(iz)$.

If $F(z)$ is analytic within and on the unit circle C, except at the singular points z_j $(j = 1, 2, \ldots, n)$ that lie interior to C, then from the residue theorem

$$
\int_0^{2\pi} f(\theta)d\theta = 2\pi i \sum_{j=1}^{n} \operatorname*{Res}_{z=z_j} F(z)
\tag{5.87}
$$

which states that the definite integral of a function $f(\theta)$ in terms of $\cos\theta$ and $\sin\theta$ can be obtained by using the residues of $F(z) = f(\theta)/(iz)$ at the poles interior to the unit circle.

Example 5.41 Let us evaluate the integral of $f(\theta) = 1/(a+b\sin\theta)$ over the interval $0 \leq \theta \leq 2\pi$, where $a > |b|$. By setting $e^{i\theta} = z$, $F(z)$ is given by

$$
F(z) = \frac{1}{iz}\frac{1}{a + b(z - 1/z)/2i} = \frac{2}{bz^2 + 2aiz - b}
$$

The denominator of $F(z)$ has zeros at $z_1 = (-a + \sqrt{a^2 - b^2})i/b$ and $z_2 = (-a - \sqrt{a^2 - b^2})i/b$ and $F(z)$ can be rewritten as

$$F(z) = \frac{2/b}{(z - z_1)(z - z_2)}$$

Since $a > |b|$, only $z_1 = (-a + \sqrt{a^2 - b^2})i/b$ is interior to the unit circle C, which is a pole of $F(z)$. From Eq. 5.78, the corresponding residue is found by writing

$$F(z) = \frac{2/b}{(z - z_1)(z - z_2)} = \frac{h(z)}{z - z_1}$$

and consequently

$$\operatorname*{Res}_{z = z_1} F(z) = h(z_1) = \frac{2/b}{z_1 - z_2} = \frac{1}{\sqrt{a^2 - b^2}i}$$

From Eq. 5.87, the definite integral is evaluated as

$$\int_0^{2\pi} \frac{1}{a + b \sin \theta} d\theta = 2\pi i \frac{1}{\sqrt{a^2 - b^2}i} = \frac{2\pi}{\sqrt{a^2 - b^2}}$$

5.5.4.2 Improper Integrals of Rational Functions

If $f(x)$ is a continuous function of the real variable x on the interval $0 \le x < \infty$, the improper integral of $f(x)$ over the semi-infinite interval $0 \le x$ is defined by

$$\int_0^\infty f(x)dx = \lim_{R \to \infty} \int_0^R f(x)dx \tag{5.88}$$

provided the limit exists. If $f(x)$ is continuous for all real x, the integral of $f(x)$ over $(-\infty, \infty)$ is defined by

$$\int_{-\infty}^\infty f(x)dx = \lim_{R_1 \to \infty} \int_{-R_1}^0 f(x)dx + \lim_{R_2 \to \infty} \int_0^{R_2} f(x)dx \tag{5.89}$$

If both limits exist, the integral converges to the value being the sum of the values of these two limits, and it follows that

$$\int_{-\infty}^\infty f(x)dx = \lim_{R \to \infty} \int_{-R}^R f(x)dx \tag{5.90}$$

It should be noted that the limit in Eq. 5.90 may exist even if the limits in Eq. 5.89 do not. For instance, when $f(x) = x$, Eq. 5.90 gives 0, whereas the integral does not converge according to Eq. 5.89. To extend the notion of the value of an improper integral, the Cauchy principal value is defined by

$$\text{P.V.} \int_{-\infty}^{\infty} f(x)dx = \lim_{R \to \infty} \int_{-R}^{R} f(x)dx \tag{5.91}$$

provided the limit exists.

Now, let us consider a rational function $f(x) = p(x)/q(x)$, where $q(x) \neq 0$ for all real x and the degree of $q(x)$ is at least two units higher than the degree of $p(x)$. Since $f(x)$ is rational, the corresponding complex function $f(z)$ has a finite number n of poles in the upper half plane. It is possible to find a real number R such that all these poles at z_j are enclosed by the contour C, which consists of the segment $-R \leq x \leq R$ of the x axis and the upper semicircle C_R of radius R, as shown in Fig. 5.5 for the case $n = 3$.

From the residue theorem, it follows that

$$\oint_C \frac{p(z)}{q(z)}dz = \int_{-R}^{R} \frac{p(x)}{q(x)}dx + \int_{C_R} \frac{p(z)}{q(z)}dz = 2\pi i \sum_{j=1}^{n} \operatorname*{Res}_{z=z_j} \frac{p(z)}{q(z)} \tag{5.92}$$

where z_j $(j = 1, 2, \ldots, n)$ are the poles of $p(z)/q(z)$ in the upper half plane. As R approaches ∞, Eq. 5.92 is rewritten as

$$\text{P.V.} \int_{-\infty}^{\infty} \frac{p(x)}{q(x)}dx = 2\pi i \sum_{j=1}^{n} \operatorname*{Res}_{z=z_j} \frac{p(z)}{q(z)} - \lim_{R \to \infty} \int_{C_R} \frac{p(z)}{q(z)}dz \tag{5.93}$$

Fig. 5.5 Simple closed contour C consisting of the line segment $-R \leq x \leq R$ and the upper semicircle C_R of radius R which encloses poles at z_j

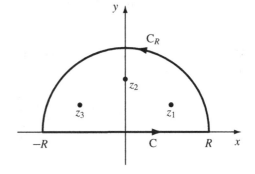

By assumption, the degree of $q(x)$ is at least two units higher than the degree of $p(x)$, and thus, along the semicircle C_R

$$\left| \frac{p(z)}{q(z)} \right| < \frac{M}{|z|^2} = \frac{M}{R^2} \tag{5.94}$$

where M is a sufficiently large constant. From the ML inequality

$$\left| \int_{C_R} \frac{p(z)}{q(z)} dz \right| < \frac{M}{R^2} \pi R = \frac{\pi M}{R} \tag{5.95}$$

As R approaches ∞, the value of the integral over C_R approaches zero, and Eq. 5.93 reduces to

$$\text{P.V.} \int_{-\infty}^{\infty} \frac{p(x)}{q(x)} dx = 2\pi i \sum_{j=1}^{n} \operatorname*{Res}_{z=z_j} \frac{p(z)}{q(z)} \tag{5.96}$$

which states that the improper integral of $p(x)/q(x)$ can be obtained by using the residues of $p(z)/q(z)$ at the poles in the upper half plane.

Example 5.42 Let us evaluate the integral of $f(x) = 1/(x^6 + 1)$ over the infinite interval $(-\infty, \infty)$. Since $z^6 + 1 = 0$ when $z = e^{k\pi i/6}$ for $k = 1, 3, 5, 7, 9, 11$, these are simple poles of $f(z) = 1/(z^6 + 1)$, between which only the poles $z = e^{k\pi i/6}$ for $k = 1, 3, 5$ lie in the upper half plane. From Eq. 5.74, the residues of $f(z)$ at $z = e^{k\pi i/6}$ are

$$\operatorname*{Res}_{z=e^{k\pi i/6}} f(z) = \frac{1}{6(e^{k\pi i/6})^5} = \frac{1}{6} e^{-5k\pi i/6}$$

where $k = 1, 3, 5$. From Eq. 5.96, the improper integral is evaluated as

$$\int_{-\infty}^{\infty} \frac{1}{x^6 + 1} dx = 2\pi i \frac{1}{6} \left[e^{-5\pi i/6} + e^{-15\pi i/6} + e^{-25\pi i/6} \right] = \frac{2\pi}{3}$$

Since $f(x) = 1/(x^6 + 1)$ is an even function, the integral over the semi-infinite interval $0 \le x$ is obtained as

$$\int_{0}^{\infty} \frac{1}{x^6 + 1} dx = \frac{1}{2} \int_{-\infty}^{\infty} \frac{1}{x^6 + 1} dx = \frac{\pi}{3}$$

5.5.4.3 Improper Integrals Involving Trigonometric Functions

Let us extend the integral method in the previous section to improper integrals of $f(x) = \sin(ax)p(x)/q(x)$ or $f(x) = \cos(ax)p(x)/q(x)$, where a is a positive constant, $q(x) \neq 0$ for all real x, and the degree of $q(x)$ is at least one unit higher than the degree of $p(x)$. The corresponding complex function is given by

$$f(z) = \frac{e^{iaz}p(z)}{q(z)} \tag{5.97}$$

which has a finite number n of poles in the upper half plane. The contour C consists of the segment $-R \leq x \leq R$ of the x axis and the upper semicircle C_R of radius R, which encloses all these poles at z_j.

From the residue theorem, it follows that

$$\oint_C \frac{e^{iaz}p(z)}{q(z)}dz = \int_{-R}^{R} \frac{e^{iax}p(x)}{q(x)}dx + \int_{C_R} \frac{e^{iaz}p(z)}{q(z)}dz = 2\pi i \sum_{j=1}^{n} \operatorname*{Res}_{z=z_j} \frac{e^{iaz}p(z)}{q(z)} \tag{5.98}$$

where z_j $(j = 1, 2, \ldots, n)$ are the poles of $e^{iaz}p(z)/q(z)$ in the upper half plane. As R approaches ∞, Eq. 5.98 is rewritten as

$$\text{P.V.} \int_{-\infty}^{\infty} \frac{e^{iax}p(x)}{q(x)}dx = 2\pi i \sum_{j=1}^{n} \operatorname*{Res}_{z=z_j} \frac{e^{iaz}p(z)}{q(z)} - \lim_{R\to\infty} \int_{C_R} \frac{e^{iaz}p(z)}{q(z)}dz \tag{5.99}$$

By assumption, the degree of $q(x)$ is at least one unit higher than the degree of $p(x)$, and thus, along the semicircle C_R

$$\left| \frac{p(z)}{q(z)} \right| < \frac{M}{|z|} = \frac{M}{R} \tag{5.100}$$

where M is a sufficiently large constant. By setting $Re^{i\theta} = z$, $dz/d\theta = iRe^{i\theta}$ and the integral over the semicircle C_R satisfies the following inequality:

$$\left| \int_{C_R} \frac{e^{iaz}p(z)}{q(z)}dz \right| \leq \frac{M}{R} \left| \int_0^{\pi} e^{iaRe^{i\theta}} iRe^{i\theta} d\theta \right| \tag{5.101}$$

Since $|e^{iaRe^{i\theta}}| = |e^{iaR(\cos\theta + i\sin\theta)}| = |e^{-aR\sin\theta}e^{iaR\cos\theta}| = e^{-aR\sin\theta}$, it follows that

$$\left| \int_0^{\pi} e^{iaRe^{i\theta}} iRe^{i\theta} d\theta \right| \leq R \int_0^{\pi} e^{-aR\sin\theta} d\theta = 2R \int_0^{\pi/2} e^{-aR\sin\theta} d\theta \tag{5.102}$$

For $0 \leq \theta \leq \pi/2$, $\sin \theta \geq 2\theta/\pi$, and thus

$$\int_0^{\pi/2} e^{-aR\sin\theta}\,d\theta \leq \int_0^{\pi/2} e^{-2aR\theta/\pi}\,d\theta = \frac{\pi}{2aR}(1 - e^{-aR}) \tag{5.103}$$

From Eqs. 5.101 through 5.103, it follows that

$$\left| \int_{C_R} \frac{e^{iaz}p(z)}{q(z)}\,dz \right| < \frac{M}{R}(2R)\frac{\pi}{2aR}(1 - e^{-aR}) = \frac{\pi M}{aR}(1 - e^{-aR}) \tag{5.104}$$

Since a is positive, the value of the integral over C_R approaches zero as R approaches ∞, and Eq. 5.99 reduces to

$$\text{P.V.} \int_{-\infty}^{\infty} \frac{e^{iax}p(x)}{q(x)}\,dx = 2\pi i \sum_{j=1}^{n} \operatorname*{Res}_{z=z_j} \frac{e^{iaz}p(z)}{q(z)} \tag{5.105}$$

Equating the real and imaginary parts of this equation gives

$$\begin{cases} \text{P.V.} \displaystyle\int_{-\infty}^{\infty} \cos(ax)\frac{p(x)}{q(x)}\,dx = -2\pi \sum_{j=1}^{n} \operatorname{Im}\left[\operatorname*{Res}_{z=z_j} \frac{e^{iaz}p(z)}{q(z)}\right] \\[4mm] \text{P.V.} \displaystyle\int_{-\infty}^{\infty} \sin(ax)\frac{p(x)}{q(x)}\,dx = 2\pi \sum_{j=1}^{n} \operatorname{Re}\left[\operatorname*{Res}_{z=z_j} \frac{e^{iaz}p(z)}{q(z)}\right] \end{cases} \tag{5.106}$$

which states that the improper integral of $\cos(ax)p(x)/q(x)$ and $\sin(ax)p(x)/q(x)$ can be obtained by using the residues of $e^{iaz}p(z)/q(z)$ at the poles in the upper half plane.

Example 5.43 Let us evaluate the integral of $f(x) = \cos(ax)/(x^2 + 1)$ over the semi-infinite interval $0 \leq x$, where a is a positive constant. Since $z^2 + 1 = 0$ when $z = \pm i$, these are simple poles of $f(z) = e^{iaz}/(z^2 + 1)$, between which only the pole $z = i$ lies in the upper half plane. From Eq. 5.74, the residue of $f(z)$ at $z = i$ is

$$\operatorname*{Res}_{z=i} f(z) = \frac{e^{-a}}{2i}$$

From Eq. 5.106, the improper integral is evaluated as

$$\int_{-\infty}^{\infty} \frac{\cos(ax)}{x^2 + 1}\,dx = -2\pi\left(-\frac{e^{-a}}{2}\right) = \pi e^{-a}$$

Since $f(x)$ is an even function, the integral over the semi-infinite interval $0 \le x$ is obtained as

$$\int_0^\infty \frac{\cos(ax)}{x^2 + 1}\,dx = \frac{\pi e^{-a}}{2}$$

5.5.4.4 Integrals with Indented Contours

If $f(z)$ has a simple pole at x_0 on the real axis, $f(z)$ has the Laurent series

$$f(z) = \underset{z=x_0}{\mathrm{Res}}\, f(z)\frac{1}{z - x_0} + g(z) \tag{5.107}$$

which is valid for $0 < |z - x_0| < R_2$ and $g(z)$ is analytic at x_0. Let us consider the integral of $f(z)$ over a positively oriented semicircle C_r, $z = x_0 + re^{i\theta}$ ($0 \le \theta \le \pi$)

$$\int_{C_r} f(z)dz = \underset{z=x_0}{\mathrm{Res}}\, f(z) \int_{C_r} \frac{1}{z - x_0}dz + \int_{C_r} g(z)dz \tag{5.108}$$

Since $dz/d\theta = ire^{i\theta}$, the integral becomes

$$\int_{C_r} f(z)dz = \underset{z=x_0}{\mathrm{Res}}\, f(z) \int_0^\pi \frac{1}{re^{i\theta}}ire^{i\theta}\,d\theta + \int_{C_r} g(z)dz$$

$$= \pi i \underset{z=x_0}{\mathrm{Res}}\, f(z) + \int_{C_r} g(z)dz \tag{5.109}$$

As $g(z)$ is continuous at x_0, $|g(z)| = |g(x_0 + re^{i\theta})|$ is bounded on C_r, $|g(z)| \le M$. From the ML inequality

$$\left| \int_{C_r} g(z)dz \right| \le M(\pi r) \tag{5.110}$$

which approaches zero as r approaches zero, and it follows that

$$\lim_{r \to 0} \int_{C_r} f(z)dz = \pi i \underset{z=x_0}{\mathrm{Res}}\, f(z) \tag{5.111}$$

Example 5.44 Let us evaluate the integral of $f(x) = x/(x^3 + 1)$ over the infinite interval $(-\infty, \infty)$. Since $z^3 + 1 = 0$ when $z = -1$ and $z = (1 \pm \sqrt{3}i)/2$, these are

Fig. 5.6 Simple closed
contour C consisting of the
upper semicircle C_R of
radius R, the upper
semicircle $-C_r$ of radius r,
and the segments of the real
axis that lie between the
semicircles

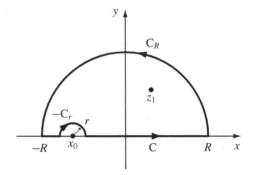

simple poles of $f(z) = z/(z^3 + 1)$, between which the pole $z_1 = (1 + \sqrt{3}i)/2$ lies
in the upper half plane and the pole $z_0 = x_0 = -1$ is on the x axis.

Since $x_0 = -1$ lies on the path of integration, the contour needs to be modified
by indenting the path at x_0, as shown in Fig. 5.6. The modified contour C consists of
the upper semicircle C_R of radius R with its center at the origin, the upper semicircle
$-C_r$ of radius r with its center at x_0, and the segments of the real axis that lie between
the semicircles. The radius R is large enough that the pole at z_1 lies under C_R and
the radius r is small enough that the pole at z_1 lies above C_r.

From the residue theorem, it follows that

$$\oint_C f(z)dz = \int_{-R}^{x_0-r} f(x)dx - \int_{C_r} f(z)dz + \int_{x_0+r}^{R} f(x)dx + \int_{C_R} f(z)dz$$

$$= 2\pi i \operatorname*{Res}_{z=z_1} f(z)$$

As shown in the previous sections, the value of the integral over C_R approaches zero
as R approaches ∞.

$$\int_{-\infty}^{x_0-r} f(x)dx - \int_{C_r} f(z)dz + \int_{x_0+r}^{\infty} f(x)dx = 2\pi i \operatorname*{Res}_{z=z_1} f(z)$$

When r approaches zero, from Eq. 5.111, it follows that

$$\text{P.V.} \int_{-\infty}^{\infty} f(x)dx = 2\pi i \operatorname*{Res}_{z=z_1} f(z) + \pi i \operatorname*{Res}_{z=x_0} f(z) \tag{5.112}$$

From Eq. 5.74, the residue of $f(z)$ at $z_1 = (1 + \sqrt{3}i)/2$ is

$$\operatorname*{Res}_{z=z_1} f(z) = \frac{(1 + \sqrt{3}i)/2}{3(1 + \sqrt{3}i)^2/4} = \frac{1 - \sqrt{3}i}{6}$$

and the residue of $f(z)$ at $z_0 = -1$ is

$$\operatorname*{Res}_{z=z_0} f(z) = \frac{(-1)}{3(-1)^2} = -\frac{1}{3}$$

From Eq. 5.112, the improper integral is evaluated as

$$\int_{-\infty}^{\infty} \frac{x}{x^3 + 1} dx = 2\pi i \frac{1 - \sqrt{3}i}{6} + \pi i \left(-\frac{1}{3}\right) = \frac{\pi}{\sqrt{3}}$$

If $f(z)$ has a finite number n of poles at z_j in the upper half plane and a finite number m of simple poles at x_j on the real axis, Eq. 5.112 is generalized in the form

$$\text{P.V.} \int_{-\infty}^{\infty} f(x)dx = 2\pi i \sum_{j=1}^{n} \operatorname*{Res}_{z=z_j} f(z) + \pi i \sum_{j=1}^{m} \operatorname*{Res}_{z=x_j} f(z) \tag{5.113}$$

When $f(x)$ is a function $p(x)/q(x)$ with the degree of $q(x)$ at least two units higher than the degree of $p(x)$, it follows that

$$\text{P.V.} \int_{-\infty}^{\infty} \frac{p(x)}{q(x)} dx = 2\pi i \sum_{j=1}^{n} \operatorname*{Res}_{z=z_j} \frac{p(z)}{q(z)} + \pi i \sum_{j=1}^{m} \operatorname*{Res}_{z=x_j} \frac{p(z)}{q(z)} \tag{5.114}$$

When $f(x)$ is a function $\sin(ax)p(x)/q(x)$ or $\cos(ax)p(x)/q(x)$, where a is a positive constant, with the degree of $q(x)$ at least one unit higher than the degree of $p(x)$, it follows that

$$\left\{ \begin{aligned} \text{P.V.} \int_{-\infty}^{\infty} \cos(ax)\frac{p(x)}{q(x)} dx &= -2\pi \sum_{j=1}^{n} \text{Im} \left[\operatorname*{Res}_{z=z_j} \frac{e^{iaz} p(z)}{q(z)} \right] \\ &\quad -\pi \sum_{j=1}^{m} \text{Im} \left[\operatorname*{Res}_{z=x_j} \frac{e^{iaz} p(z)}{q(z)} \right] \\ \text{P.V.} \int_{-\infty}^{\infty} \sin(ax)\frac{p(x)}{q(x)} dx &= 2\pi \sum_{j=1}^{n} \text{Re} \left[\operatorname*{Res}_{z=z_j} \frac{e^{iaz} p(z)}{q(z)} \right] \\ &\quad +\pi \sum_{j=1}^{m} \text{Re} \left[\operatorname*{Res}_{z=x_j} \frac{e^{iaz} p(z)}{q(z)} \right] \end{aligned} \right. \tag{5.115}$$

Example 5.45 Let us evaluate the integral of $f(x) = 1/(x^4 - 1)$ over the infinite interval $(-\infty, \infty)$. Since $z^4 - 1 = 0$ when $z = \pm i$ and $z = \pm 1$, these are simple poles of $f(z) = 1/(z^4 - 1)$, between which the pole $z = i$ lies in the upper half plane and the poles $z = \pm 1$ are on the x axis. From Eq. 5.74, the residue of $f(z)$ at $z = i$ is

$$\operatorname*{Res}_{z=i} f(z) = \frac{1}{4i^3} = \frac{i}{4}$$

and the residues of $f(z)$ at $z = \pm 1$ are

$$\operatorname*{Res}_{z=\pm 1} f(z) = \frac{1}{4(\pm 1)^3} = \pm \frac{1}{4}$$

From Eq. 5.114, the improper integral is evaluated as

$$\int_{-\infty}^{\infty} \frac{1}{x^4 - 1} dx = 2\pi i \frac{i}{4} + \pi i \left(\frac{1}{4} - \frac{1}{4} \right) = -\frac{\pi}{2}$$

Example 5.46 Let us evaluate the integral of $f(x) = \sin x / x$ over the semi-infinite interval $0 \le x$. The corresponding complex function $f(z) = e^{iz}/z$ has a simple pole at $z = 0$ on the x axis. From Eq. 5.74, the residue of $f(z)$ at $z = 0$ is

$$\operatorname*{Res}_{z=0} f(z) = e^0 = 1$$

From Eq. 5.115, the improper integral is evaluated as

$$\int_{-\infty}^{\infty} \frac{\sin x}{x} dx = 2\pi(0) + \pi(1) = \pi$$

Since $f(x)$ is an even function, the integral over the semi-infinite interval $0 \le x$ is obtained as

$$\int_{0}^{\infty} \frac{\sin x}{x} dx = \frac{\pi}{2}$$

Example 5.47 Let us evaluate the integral of $f(x) = \sin(\pi x)/x(1 - x^2)$ over the semi-infinite interval $0 \le x$. The corresponding complex function $f(z) = e^{i\pi z}/z(1 - z^2)$ has simple poles at $z = 0, \pm 1$ on the x axis. From Eq. 5.74, the residue of $f(z)$ at $z = 0$ is

$$\operatorname*{Res}_{z=0} f(z) = e^0 = 1$$

and the residues of $f(z)$ at $z = \pm 1$ are

$$\operatorname*{Res}_{z=\pm 1} f(z) = \frac{e^{\pm \pi i}}{1 - 3(\pm 1)^2} = -\frac{e^{\pm \pi i}}{2} = \frac{1}{2}$$

From Eq. 5.115, the improper integral is evaluated as

$$\int_{-\infty}^{\infty} \frac{\sin \pi x}{x(1-x^2)} dx = 2\pi(0) + \pi\left(1 + \frac{1}{2} + \frac{1}{2}\right) = 2\pi$$

Since $f(x)$ is an even function, the integral over the semi-infinite interval $0 \le x$ is obtained as

$$\int_0^{\infty} \frac{\sin \pi x}{x(1-x^2)} dx = \pi$$

5.5.4.5 Improper Integrals Involving Multi-Valued Functions

Let us consider a function $f(x) = x^{-a} p(x)/q(x)$, where a is a real constant with $0 < a < 1$, $q(x) \ne 0$ for $x \ge 0$ and the degree of $q(x)$ is at least one unit higher than the degree of $p(x)$. The corresponding complex function is given by $f(z) = z^{-a} p(z)/q(z)$, which has a finite number n of poles. The complex term z^a is multi-valued since

$$z^a = e^{a \ln z} = e^{a(\ln |z| + i\theta)} = |z|^a e^{ia\theta} \tag{5.116}$$

where $\theta = \arg z$.

If the positive real half-axis is chosen to be the branch cut, $\arg z$ takes the principal value $0 \le \text{Arg } z < 2\pi$. The contour C consists of the portions of the circle C_R of radius R and the circle C_r of radius r and the horizontal segments joining them, as shown in Fig. 5.7. The radius R is large enough and the radius r is small enough that all n poles lie inside C.

Fig. 5.7 Simple closed contour C consisting of the portions of the circle C_R of radius R and the circle C_r of radius r and the horizontal segments joining them

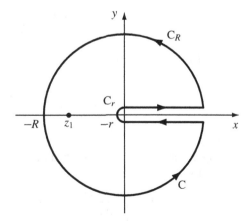

When r approaches zero, the upper and lower horizontal segments of C approaches the line segment from 0 to R along the upper side of the x axis, $z = xe^0 = x$, and the line segment from R to 0 along the lower side of the x axis, $z = xe^{2\pi i}$, respectively. Then, the integral over C becomes

$$\lim_{r \to 0} \oint_C \frac{p(z)}{z^a q(z)} dz = \int_0^R \frac{p(x)}{x^a q(x)} dx + \int_R^0 \frac{p(x)}{x^a e^{2\pi a i} q(x)} dx + \int_{C_R} \frac{p(z)}{z^a q(z)} dz$$

$$= (1 - e^{-2\pi a i}) \int_0^R \frac{p(x)}{x^a q(x)} dx + \int_{C_R} \frac{p(z)}{z^a q(z)} dz \qquad (5.117)$$

From the residue theorem, it follows that

$$(1 - e^{-2\pi a i}) \int_0^R \frac{p(x)}{x^a q(x)} dx + \int_{C_R} \frac{p(z)}{z^a q(z)} dz = 2\pi i \sum_{j=1}^n \operatorname*{Res}_{z=z_j} \frac{p(z)}{z^a q(z)} \qquad (5.118)$$

where z_j $(j = 1, 2, \ldots, n)$ are the poles of $z^{-a} p(z)/q(z)$. As R approaches ∞, Eq. 5.118 is rewritten as

$$(1 - e^{-2\pi a i}) \, \text{P.V.} \int_0^\infty \frac{p(x)}{x^a q(x)} dx = 2\pi i \sum_{j=1}^n \operatorname*{Res}_{z=z_j} \frac{p(z)}{z^a q(z)}$$

$$- \lim_{R \to \infty} \int_{C_R} \frac{p(z)}{z^a q(z)} dz \qquad (5.119)$$

By assumption, the degree of $q(x)$ is at least one unit higher than the degree of $p(x)$, and thus, along the circle C_R

$$\left| \frac{p(z)}{z^a q(z)} \right| < \frac{M}{|z|^{1+a}} = \frac{M}{R^{1+a}} \qquad (5.120)$$

where M is a sufficiently large constant. From the ML inequality

$$\left| \int_{C_R} \frac{p(z)}{z^a q(z)} dz \right| < \frac{M}{R^{1+a}} 2\pi R = \frac{2\pi M}{R^a} \qquad (5.121)$$

As R approaches ∞, the value of the integral over C_R approaches zero, and Eq. 5.119 reduces to

$$\text{P.V.} \int_0^\infty \frac{p(x)}{x^a q(x)} dx = \frac{2\pi i}{1 - e^{-2\pi a i}} \sum_{j=1}^n \operatorname*{Res}_{z=z_j} \frac{p(z)}{z^a q(z)} \qquad (5.122)$$

which states that the improper integral of $x^{-a}p(x)/q(x)$ can be obtained by using the residues of $z^{-a}p(z)/q(z)$ at the nonzero poles.

Example 5.48 Let us evaluate the integral of $f(x) = x^{-a}/(x+1)$ over the semi-infinite interval $0 \leq x$, where $0 < a < 1$. The corresponding complex function $f(z) = z^{-a}/(z+1)$ has a nonzero pole at $z = -1$. From Eq. 5.78, the residue of $f(z)$ at $z = -1$ is

$$\mathop{\mathrm{Res}}_{z=-1} f(z) = (-1)^{-a} = e^{-\pi a i}$$

From Eq. 5.122, the improper integral is evaluated as

$$\int_0^\infty \frac{x^{-a}}{x+1} dx = \frac{2\pi i}{1 - e^{-2\pi a i}} e^{-\pi a i} = \frac{2\pi i}{e^{\pi a i} - e^{-\pi a i}} = \frac{\pi}{\sin a\pi}$$

5.6 Applications of Laurent Series

Laurent series representations of analytic functions are of use to derive approximate solutions to engineering problems, when the corresponding closed-form solutions are not available.

5.6.1 Flow Through a Distorted Circle

Uniform flow through an equipotential circle is considered in Sect. 2.6.5, where the complex potential (Eq. 2.166) is obtained by the superposition of uniform flow and a dipole, that is,

$$\Omega(z) = -z + \frac{1}{z} \tag{5.123}$$

which is the Laurent series about the singular point $z_0 = 0$ located outside the domain of interest. As z approaches ∞, $\Omega(z)$ tends to uniform flow $\Omega = -z$.

Suppose that the equipotential object is not a perfect circle and its shape is slightly distorted according to

$$r = 1 + \varepsilon \cos^2 \theta \tag{5.124}$$

where ε is a small quantity, as shown in Fig. 5.8.

The complex potential of uniform flow through the distorted circle can be represented by a generalized form of Eq. 5.123 as

$$\Omega(z) = -z + \frac{a_{-1}}{z} + \frac{a_{-2}}{z^2} + \cdots + \frac{a_{-n}}{z^n} + \cdots \tag{5.125}$$

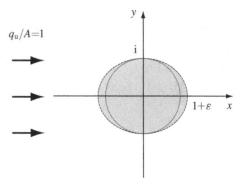

Fig. 5.8 A distorted circle
with equipotential
circumference in uniform
flow

where the coefficients a_{-n} are real numbers. It must be noted that, as is the case with
the perfect circle, $\Omega(z)$ tends to uniform flow $\Omega = -z$ as z approaches ∞.

The circumference of the distorted circle is an equipotential line, which leads to
the following boundary condition:

$$\Phi(r, \theta) = \Phi(1 + \varepsilon \cos^2 \theta, \theta) = 0 \qquad (5.126)$$

By using the exponential form $z = re^{i\theta}$ in Eq. 5.125, the velocity potential is found
to be

$$\Phi(r, \theta) = -r \cos \theta + a_{-1} \frac{\cos \theta}{r} + a_{-2} \frac{\cos 2\theta}{r^2} + \cdots + a_{-n} \frac{\cos n\theta}{r^n} + \cdots$$

$$= -r \cos \theta + \sum_{n=1}^{\infty} a_{-n} \frac{\cos n\theta}{r^n} \qquad (5.127)$$

where the coefficients a_{-n} are to be determined by using the boundary condition,
Eq. 5.126.

5.6.1.1 Perturbation Method

The perturbation method is a collection of techniques for analyzing the global behavior of the solution for a differential equation which involve a perturbation parameter ε.
If the solution can be expressed as a convergent series in powers of ε, the problem
is referred to as a regular perturbation problem. In such a problem, a perturbation
series often provides ample accuracy with a few terms by virtue of its asymptotic
nature.

Let us assume that the solution can be expressed as a perturbation series in powers
of ε

$$\Phi(r, \theta) = \Phi_0(r, \theta) + \varepsilon \Phi_1(r, \theta) + \varepsilon^2 \Phi_2(r, \theta) + \cdots + \varepsilon^n \Phi_n(r, \theta) + \cdots \quad (5.128)$$

where $\Phi_n(r, \theta)$ is the nth-order perturbation solution. The zeroth-order solution corresponds to $\varepsilon = 0$ and is the real part of Eq. 5.123, that is,

$$\Phi_0(r, \theta) = -r\cos\theta + \frac{\cos\theta}{r} \tag{5.129}$$

For the higher order solutions, the boundary condition Eq. 5.126 is applied. Substituting Eq. 5.128 into Eq. 5.126 gives

$$\Phi_0(1 + \varepsilon\cos^2\theta, \theta) + \varepsilon\Phi_1(1 + \varepsilon\cos^2\theta, \theta)$$
$$+ \varepsilon^2\Phi_2(1 + \varepsilon\cos^2\theta, \theta) + \cdots = 0 \tag{5.130}$$

Expanding $\Phi_n(1 + \varepsilon\cos^2\theta, \theta)$ in a Taylor series about $r = 1$ yields

$$\Phi_n(1 + \varepsilon\cos^2\theta, \theta) = \sum_{m=0}^{\infty} \frac{\Phi_n^{(m)}(1, \theta)}{m!} (\varepsilon\cos^2\theta)^m$$

$$= \Phi_n(1, \theta) + \varepsilon\cos^2\theta\frac{\partial\Phi_n}{\partial r}(1, \theta) + \frac{\varepsilon^2\cos^4\theta}{2!}\frac{\partial^2\Phi_n}{\partial r^2}(1, \theta)$$

$$+ \frac{\varepsilon^3\cos^6\theta}{3!}\frac{\partial^3\Phi_n}{\partial r^3}(1, \theta) + \cdots \tag{5.131}$$

Substituting Eq. 5.131 into Eq. 5.130 and arranging the coefficients of equal powers of ε gives

$$\Phi_0(1, \theta) + \varepsilon\left[\cos^2\theta\frac{\partial\Phi_0}{\partial r}(1, \theta) + \Phi_1(1, \theta)\right]$$
$$+ \varepsilon^2\left[\frac{\cos^4\theta}{2!}\frac{\partial^2\Phi_0}{\partial r^2}(1, \theta) + \cos^2\theta\frac{\partial\Phi_1}{\partial r}(1, \theta) + \Phi_2(1, \theta)\right]$$
$$+ \varepsilon^3\left[\frac{\cos^6\theta}{3!}\frac{\partial^3\Phi_0}{\partial r^3}(1, \theta) + \frac{\cos^4\theta}{2!}\frac{\partial^2\Phi_1}{\partial r^2}(1, \theta)\right.$$
$$\left. + \cos^2\theta\frac{\partial\Phi_2}{\partial r}(1, \theta) + \Phi_3(1, \theta)\right] + \cdots = 0 \tag{5.132}$$

The first-order solution satisfies the boundary condition

$$\cos^2\theta\frac{\partial\Phi_0}{\partial r}(1, \theta) + \Phi_1(1, \theta) = 0 \tag{5.133}$$

which results in

$$\Phi_1(1, \theta) = -\cos^2\theta(-\cos\theta - \cos\theta) = 2\cos^3\theta = \frac{3}{2}\cos\theta + \frac{1}{2}\cos 3\theta \tag{5.134}$$

Since $\Phi(r, \theta)$ takes the form given by Eq. 5.127, $\Phi_1(r, \theta)$ is obtained as

$$\Phi_1(r, \theta) = \frac{3}{2}\frac{\cos \theta}{r} + \frac{1}{2}\frac{\cos 3\theta}{r^3} \tag{5.135}$$

In a similar way, the higher order $(n > 1)$ solutions $\Phi_n(r, \theta)$ can be obtained by solving the corresponding boundary condition repeatedly with known solutions $\Phi_j(r, \theta)$ $(j < n)$ from the previous perturbation. The second-order and third-order solutions are found to be

$$\begin{cases} \Phi_2(r, \theta) = \frac{7}{8}\frac{\cos \theta}{r} + \frac{13}{16}\frac{\cos 3\theta}{r^3} + \frac{5}{16}\frac{\cos 5\theta}{r^5} \\ \Phi_3(r, \theta) = -\frac{1}{16}\frac{\cos \theta}{r} + \frac{9}{16}\frac{\cos 3\theta}{r^3} + \frac{21}{32}\frac{\cos 5\theta}{r^5} + \frac{7}{32}\frac{\cos 7\theta}{r^7} \end{cases} \tag{5.136}$$

Thus, the complete third-order solution is obtained:

$$\tilde{\Omega}(z) = -z + \frac{1}{z} + \varepsilon \left(\frac{3}{2}\frac{1}{z} + \frac{1}{2}\frac{1}{z^3}\right) + \varepsilon^2 \left(\frac{7}{8}\frac{1}{z} + \frac{13}{16}\frac{1}{z^3} + \frac{5}{16}\frac{1}{z^5}\right)$$
$$+ \varepsilon^3 \left(-\frac{1}{16}\frac{1}{z} + \frac{9}{16}\frac{1}{z^3} + \frac{21}{32}\frac{1}{z^5} + \frac{7}{32}\frac{1}{z^7}\right) \tag{5.137}$$

and the coefficients of the Laurent series Eq. 5.125 are determined as

$$\begin{cases} a_{-1} = 1 + \frac{3}{2}\varepsilon + \frac{7}{8}\varepsilon^2 - \frac{1}{16}\varepsilon^3 \\ a_{-3} = \frac{1}{2}\varepsilon + \frac{13}{16}\varepsilon^2 + \frac{9}{16}\varepsilon^3 \\ a_{-5} = \frac{5}{16}\varepsilon^2 + \frac{21}{32}\varepsilon^3 \\ a_{-7} = \frac{7}{32}\varepsilon^3 \end{cases} \tag{5.138}$$

Figure 5.9 shows the equipotential lines and streamlines for $\varepsilon = 0.2$. Flow through the distorted circle is appropriately modeled by the Laurent series representation.

5.6.2 Flow Around a Source-Sink Pair

Let us consider flow around a source-sink pair; a source with a flow rate $q_{w1}/h = 2\pi$ at $(1, 0)$ and a sink with a flow rate $q_{w2}/h = -2\pi$ at $(-1, 0)$. By use of superposition, the complex potential is obtained as

$$\Omega(z) = -\frac{q_{w1}}{2\pi h}\ln(z - 1) - \frac{q_{w2}}{2\pi h}\ln(z + 1) = \ln\frac{z + 1}{z - 1} \tag{5.139}$$

(a)   **(b)**

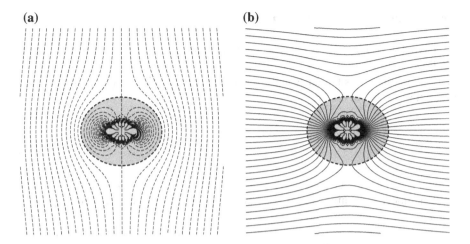

Fig. 5.9 Uniform flow through a distorted circle. **a** Equipotential lines. **b** Streamlines

(a)   **(b)**

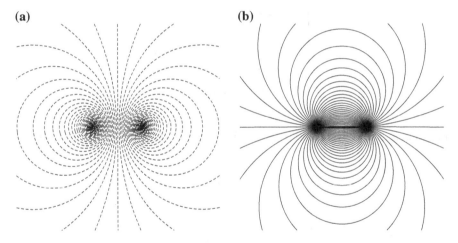

Fig. 5.10 Flow around a source-sink pair. **a** Equipotential lines. **b** Streamlines

Figure 5.10 shows the corresponding equipotential lines and streamlines.

Now, suppose that the domain of interest is exterior to a circle of radius R with its center at the origin, $|z| > R$, and that the cause of flow, that is, the existence of a source-sink pair, is not known. Hence, the solution in a closed form (Eq. 5.139) is not available. Instead, flow is known to vanish at points far from the circle and the velocity potentials at m evenly spaced observation points along the circumference of the circle are given. Figure 5.11a, b show the circles and observation points ($m = 10$) for $R = 2$ and $R = 1.2$, respectively.

5.6.2.1 Laurent Series with a Single Pole

Let us find the Laurent series representation for flow around a circle of $R = 2$. Since the flow vanishes as $z \to \infty$, the analytic part of the Laurent series must be zero. Here, the complex potential is assumed to be represented by a partial sum of the Laurent series about the origin:

$$\tilde{\Omega}(z) = \sum_{n=1}^{n_{\mathrm{L}}} \frac{a_{-n}}{z^n} \tag{5.140}$$

where n_{L} is the number of negative power terms and a_{-n} are real numbers. For a_{-n} to be uniquely determined, n_{L} is set equal to m, the number of observation points. Equation 5.140 must be valid for $|z| \geq R = 2$.

The coefficients a_{-n} can be determined from conditions applied at observation points on the circle. For 10 observation points ($m = n_{\mathrm{L}} = 10$), as shown in Fig. 5.11a, a_{-n} for $n = 1, 2, \ldots, 10$ are obtained as

$$\begin{cases} a_{-1} = 1.9996 = 2/1.0002 \\ a_{-3} = 0.66636 = 2/3.0014 \\ a_{-5} = 0.39961 = 2/5.0049 \\ a_{-7} = 0.28309 = 2/7.0651 \\ a_{-9} = 0.17661 = 2/11.324 \end{cases} \tag{5.141}$$

and $a_{-2} = a_{-4} = a_{-6} = a_{-8} = a_{-10} = 0$. For the exact solution, $\ln[(z+1)/(z-1)]$, the Laurent series is given by Eq. 5.36, which is valid for $|z| > 1$, and the

(a)   **(b)**

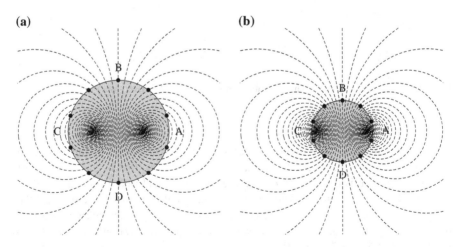

Fig. 5.11 Domain of interest exterior to a circle of radius R and 10 observation points on the circle. **a** Circle of radius $R = 2$. **b** Circle of radius $R = 1.2$

coefficients are

$$\begin{cases} a_{-n} = 2/n & \text{(if } n \text{ is odd)} \\ a_{-n} = 0 & \text{(if } n \text{ is even)} \end{cases} \qquad (5.142)$$

The solutions a_{-n} (Eq. 5.141) agree well with the exact values (Eq. 5.142) except a_{-9}, with which the disregarded terms ($n > 11$) are taken into account.

To verify the solution, approximate values of $\tilde{\Omega}(z)$ are compared with the exact solution (Eq. 5.139) at 200 evenly spaced points along the circumference of the circle. The maximum of the absolute error (AE) $|\tilde{\Omega}(z) - \Omega(z)|$ is 4.51×10^{-4} and the mean absolute error (MAE) is quite small (2.01×10^{-4}). Figure 5.12 shows the equipotential lines and streamlines by using $\tilde{\Omega}(z)$. Flow around the circle of $R = 2$ is appropriately modeled with the approximate solution. For the case of $R = 2$, the partial sum of Laurent series ($n_L = 10$) about a single pole is found to be accurate enough for practical purposes.

In a similar way, the Laurent series representation for flow around the circle of $R = 1.2$ is considered. Equation 5.140 is again used, which must be valid for $|z| \geq 1.2$. For 10 observation points, as shown in Fig. 5.11b, a_{-n} are obtained as

$$\begin{cases} a_{-1} = 1.9445 = 2/1.0286 \\ a_{-3} = 0.61531 = 2/3.2504 \\ a_{-5} = 0.34438 = 2/5.8075 \\ a_{-7} = 0.20924 = 2/9.5584 \\ a_{-9} = 0.084959 = 2/23.541 \end{cases} \qquad (5.143)$$

(a)   **(b)**

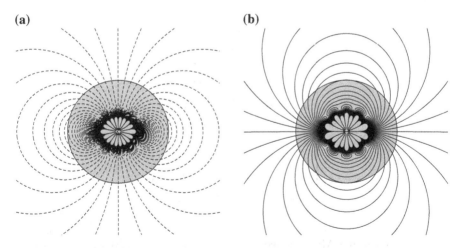

Fig. 5.12 Flow around a circle of $R = 2$ evaluated by the partial sum of Laurent series with a single pole. **a** Equipotential lines. **b** Streamlines

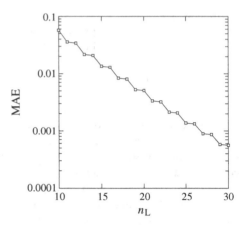

Fig. 5.13 Mean absolute error along the circumference of the circle of $R = 1.2$ for different values of n_L

and $a_{-2} = a_{-4} = a_{-6} = a_{-8} = a_{-10} = 0$. Compared with the previous case of $R = 2$, the agreement with the exact values (Eq. 5.142) is aggravated. Consequently, the maximum of the AE (0.208) and the MAE (5.75×10^{-2}) are much higher, and the Laurent series representation with $n_L = 10$ becomes erroneous for the case of $R = 1.2$.

Although it requires more observation points, a possible way to improve the accuracy is to increase the number of negative terms n_L of the Laurent series. Figure 5.13 shows the MAE characteristics with respect to n_L. As n_L increases, the MAE indeed decreases. Note that the Laurent series with $n_L = 2n$ yields little improvement over that with $n_L = 2n - 1$, since a_{-2n} is zero.

For $n_L = 20$, and thus $m = 20$, a_{-n} are obtained as

$$\begin{cases} a_{-1} = 1.9951 = 2/1.0025 \\ a_{-3} = 0.66212 = 2/3.0206 \\ a_{-5} = 0.39568 = 2/5.0546 \\ a_{-7} = 0.28142 = 2/7.1068 \\ a_{-9} = 0.21757 = 2/9.1924 \\ a_{-11} = 0.17602 = 2/11.362 \\ a_{-13} = 0.14516 = 2/13.778 \\ a_{-15} = 0.11787 = 2/16.968 \\ a_{-17} = 0.086567 = 2/23.103 \\ a_{-19} = 0.038240 = 2/52.301 \end{cases} \qquad (5.144)$$

and $a_{-2} = a_{-4} = a_{-6} = a_{-8} = a_{-10} = a_{-12} = a_{-14} = a_{-16} = a_{-18} = a_{-20} = 0$. The maximum of the AE is reduced to 2.24×10^{-2} and the MAE is reduced to 5.09×10^{-3}. However, the solutions for a_{-n} are less accurate than the case for the circle of $R = 2$ with $n_L = 10$ (Eq. 5.141). In particular, the coefficients for higher order terms are erroneous. For smaller values of R, increasing n_L alone cannot essentially overcome the modeling difficulty.

5.6.2.2 Laurent Series with Multiple Poles

An alternative way to improve the accuracy of a Laurent series representation is to use multiple poles, resulting in several partial sums of different Laurent series expansions. To avoid introducing a singularity in the domain of interest, poles must be established at locations exterior to the domain. By locating a pole near the region where flow exhibits singular behavior, the modeling accuracy can be improved with a lower order Laurent series representation.

Figure 5.14 shows the AE profiles along the circumference of the circle of $R = 1.2$. For the Laurent series expansion with $n_L = 10$ about a single pole at the origin, the maximum value of 0.208 is found at the points A and C on the circle, around which singularities exist as seen in Fig. 5.11b. Hence, allocating poles near the points A and C could be a good modeling strategy.

The complex potential is assumed to be represented by the partial sums of Laurent series about the origin and the first-order poles at 0.9 and -0.9, which are exterior to the domain of interest.

$$\tilde{\Omega}(z) = \sum_{n=1}^{n_L} \frac{a_{-n}}{z^n} + \frac{b_1}{z - 0.9} + \frac{b_2}{z + 0.9} \qquad (5.145)$$

where a_{-n}, b_1, and b_2 are real numbers. For a_{-n}, b_1, and b_2 to be uniquely determined, n_L is set equal to $m - 2$. Equation 5.145 must be valid for $|z| \geq 1.2$.

For 10 observation points ($m = 10$), as shown in Fig. 5.11b, a_{-n}, b_1, and b_2 are obtained as

Fig. 5.14 Absolute error along the circumference of the circle of $R = 1.2$; truncated Laurent series expansions with a single pole and multiple poles

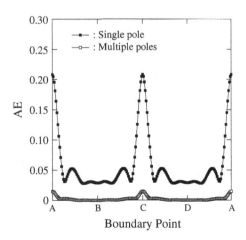

$$\begin{cases} a_{-1} = 1.4960 \\ a_{-3} = 0.25641 \\ a_{-5} = 0.065069 \\ a_{-7} = 0.012171 \end{cases}$$

$$\begin{cases} b_1 = 0.25168 \\ b_2 = 0.25168 \end{cases}$$

(5.146)

and $a_{-2} = a_{-4} = a_{-6} = a_{-8} = 0$. Since the series representation is different from the Laurent series expansion about the origin (Eq. 5.36), the solutions for a_{-n} are not necessarily equal to the solutions given by Eq. 5.142.

The AE profile along the circumference of the circle is shown in Fig. 5.14. With the same number of observation points, $m = 10$, the Laurent series representation with multiple poles yields a much smaller magnitude of error than that with a single pole. Compared even with the results for an expansion of $n_L = 20$ with a single pole, the maximum of the AE is reduced from 2.24×10^{-2} to 1.46×10^{-2} and the MAE is reduced from 5.09×10^{-3} to 2.23×10^{-3}. A series representation with multiple poles shows its ability to provide a better representation with fewer terms than a series with a single pole.

Figure 5.15 shows the equipotential lines and streamlines by using $\tilde{\Omega}(z)$, given by Eqs. 5.145 and 5.146. Flow around the circle of $R = 1.2$ is appropriately modeled with the partial sums of Laurent series with multiple poles.

(a)   **(b)**

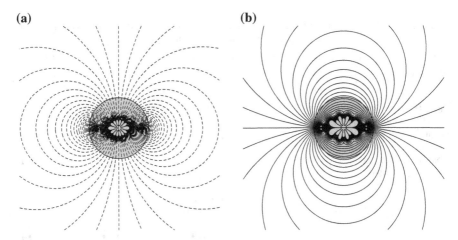

Fig. 5.15 Flow around a circle of $R = 1.2$ evaluated by the partial sums of Laurent series with multiple poles. **a** Equipotential lines. **b** Streamlines

5.7 Implicit Singularity Programming

In Task 4-3, uniform flow through an infinite conductivity fracture is modeled by the composition of conformal mappings (Eq. 3.52). For arbitrary flow through a finite conductivity fracture, a closed-form solution is generally not available and a versatile solution method is required. Here, a generalized solution for flow through a finite conductivity fracture is derived by using a Laurent series, which is then coupled with the CVBEM by the principle of superposition. The numerical procedure is called singularity programming.

5.7.1 Flow Through a Finite Conductivity Fracture

Let us consider a fracture with a finite permeability k_f and an aperture w_f with end points z_1 and z_2, as shown in Fig. 5.16. According to Eq. 2.158, the difference between the values of Ψ on the $-$ and $+$ sides of the fracture at z is equal to the flow rate per unit thickness (q_f / h) inside the fracture

$$\frac{q_f}{h} = w_f V_f = -\Delta\Psi = \Psi(z^-) - \Psi(z^+) \tag{5.147}$$

where V_f is the velocity along the fracture, defined by Darcy's law as

$$V_f = -\frac{k_f}{\mu}\frac{dp}{ds}(z) = -\frac{k_f}{k}\frac{d\Phi}{ds}(z) \tag{5.148}$$

where k is the permeability of the medium and s the coordinate along the fracture.

Combining Eq. 5.147 with Eq. 5.148 yields the boundary condition in terms of the stream function:

$$\Psi(z^+) - \Psi(z^-) = \frac{k_f w_f}{k}\frac{d\Phi}{ds}(z) \tag{5.149}$$

Fig. 5.16 A fracture with end points z_1 and z_2

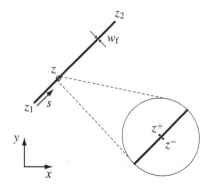

In a mathematical sense, the fracture aperture w_f is used to compute $k_k w_f / k$ in Eq. 5.149, but in a geometrical sense, w_f is negligible. Hence, the values of Φ on the $+$ and $-$ sides of the fracture must be equal to each other:

$$\Phi(z^+) - \Phi(z^-) = 0 \qquad (5.150)$$

Equations 5.149 and 5.150 are the boundary conditions along the fracture.

5.7.2 Singular Solution of Fractures

As observed in Motivating Problem 4, flow around a thin object exhibits singular behavior near the edges. To derive a singular solution that satisfies Eqs. 5.149 and 5.150, the complex potentials on both sides of the fracture must be evaluated. The following two-step conformal mapping is of use to this end:

$$\begin{cases} Z(z) = \dfrac{2z - (z_1 + z_2)}{z_2 - z_1} \\ \chi(Z) = Z + (Z^2 - 1)^{1/2} \end{cases} \qquad (5.151)$$

Figure 5.17 illustrates the mapping. The function $Z(z)$ maps the fracture with end points z_1 and z_2 onto the line segment with end points $Z_1 = Z(z_1) = -1$ and $Z_2 = Z(z_2) = 1$. The function $\chi(Z)$ is the inverse Joukowski transformation that maps the line segment $[-1, 1]$ onto the unit circle $|\chi(Z)| = 1$. It follows from the mapping function $\chi(z)$ that points (z^+ and z^-) on opposite sides of the fracture in the z plane are mapped onto points (χ^+ and χ^-) of the unit circle in the χ plane that are each other's complex conjugate.

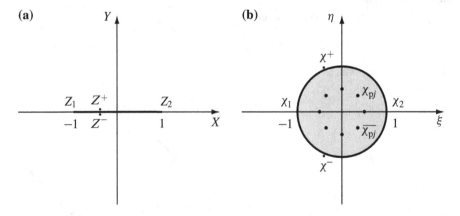

Fig. 5.17 Mapping of a fracture. **a** Line segment $[-1, 1]$ in the Z plane. **b** Unit circle in the χ plane

The fracture singular solution Ω_f that satisfies Eqs. 5.149 and 5.150 can be defined such that its real part Φ_f is continuous and its imaginary part Ψ_f is discontinuous across the fracture. In the χ plane, this can be written as

$$\begin{cases} \Phi_f(\chi^+) - \Phi_f(\chi^-) = \text{Re}[\Omega_f(\chi^+) - \Omega_f(\overline{\chi^+})] = 0 \\ \Psi_f(\chi^+) - \Psi_f(\chi^-) = \text{Im}[\Omega_f(\chi^+) - \Omega_f(\overline{\chi^+})] \neq 0 \end{cases} \tag{5.152}$$

where $\chi^- = \overline{\chi^+}$ is used. These conditions can be met if the function Ω_f is chosen such that

$$\Omega_f(\chi) = \overline{\Omega_f(\overline{\chi})} \tag{5.153}$$

Equation 5.153 states that the real part of $\Omega_f(\chi^+)$ is equal to that of $\Omega_f(\overline{\chi^+})$ and that the imaginary parts of $\Omega_f(\chi^+)$ and $\Omega_f(\overline{\chi^+})$ are opposite in sign. In addition, Ω_f must be analytic outside the fracture, and at points far from the fracture, $\chi \to \infty$, it must behave as

$$\lim_{\chi \to \infty} \Omega_f(\chi) = 0 \tag{5.154}$$

that is, the influence of the fracture vanishes at ∞.

Since the boundary condition in terms of Ψ, Eq. 5.149, is highly problem-specific, it is impossible to generalize a closed-form solution for Eqs. 5.153 and 5.154. As observed in Sect. 5.6, when a closed-form solution is not available, partial sums of Laurent series can be used to derive an approximate solution. Such a solution is versatile, since the coefficients can be determined so that the solution satisfies the individual boundary conditions.

In particular, Laurent series with multiple poles provide a robust and accurate solution for flow exhibiting singular behavior, as discussed in Sect. 5.6.2. In Eq. 5.145, a pole of order n_L at the origin and first-order poles on the real axis are used to construct the solution. To improve the robustness, distribution of additional poles is considered. Here, the fracture singular solution for Eqs. 5.153 and 5.154 is defined as a combination of partial sums of different Laurent series expansions with multiple poles given by

$$\Omega_f(\chi) = \sum_{n=1}^{n_L} \frac{a_{-n}}{\chi^n} + \sum_{j=1}^{n_p} \left(\frac{\beta_j \chi_{pj}}{\chi - \chi_{pj}} + \frac{\overline{\beta_j \chi_{pj}}}{\chi - \overline{\chi_{pj}}} \right) \tag{5.155}$$

where $\beta_j = b_j + i c_j$ and a_{-n}, b_j, and c_j are real numbers. The first expansion is the Laurent series of order n_L about the origin and the remaining expansions are the first-order Laurent series about n_p complex conjugate pairs of poles located at χ_{pj} and $\overline{\chi_{pj}}$, as shown in Fig. 5.17b. Note that the poles lie interior to the unit circle, for instance $|\chi_{pj}| = |\overline{\chi_{pj}}| = 0.5$, and thus, exterior to the flow domain.

The Laurent series about the origin equally affects the entire boundary of the fracture, while the latter ones individually affect the boundary portion close to the corresponding pole. Hence, the combination of these Laurent series yields a robust representation of both global and local (singular) flow behavior around the fracture.

5.7.3 Procedure and Formulation

In general, a complete solution Ω can be interpreted as the sum of non-singular and singular solutions, Ω^{ns} and Ω^s, that is,

$$\Omega = \Omega^{ns} + \Omega^s \tag{5.156}$$

where the principle of superposition is used. The singularity programming[1] approach utilizes Eq. 5.156 and allows us to treat the individual solutions separately.

The procedure of singularity programming is a sequence of desuperposition and superposition of singularities. By use of desuperposition, the non-singular solution can be decomposed from the complete solution: $\Omega^{ns} = \Omega - \Omega^s$. Since the non-singular component is analytic everywhere within and on a problem boundary, Cauchy's integral formula holds; thus, Ω^{ns} can be determined by the CVBEM. Once Ω^{ns} is obtained, the complete solution is recovered by use of superposition: $\Omega = \Omega^{ns} + \Omega^s$.

5.7.3.1 Implicit Scheme

In a boundary value problem containing a fracture, the singular component Ω^s is produced by Ω_f, and the approximate solution becomes

$$\tilde{\Omega}(z) = \tilde{\Omega}^{ns}(z) + \Omega_f(z) \tag{5.157}$$

where the non-singular solution $\tilde{\Omega}^{ns}(z)$ is given by Eq. 4.56 and the fracture singular solution $\Omega_f(z)$ is given by Eq. 5.155. Unknown variables to be determined are the nodal values in the non-singular solution $\tilde{\Omega}^{ns}$ and the fracture parameters involved in the fracture singular solution Ω_f.

In the CVBEM (Formulation III), the n_b-boundary discretization yields $2n_b$ non-singular nodal values: Φ_k^{ns} and Ψ_k^{ns} ($k = 1, 2, \ldots, n_b$), which are unknown. As for the fracture singular solution, when two of the poles lie on the real axis, as shown in Fig. 5.17b, there follows $n_L + 2n_p - 2$ unknown fracture parameters: a_{-n} ($n = 1, 2, \ldots, n_L$), b_j ($j = 1, 2, \ldots, n_p$), and c_j ($j = 2, \ldots, n_p - 1$). Thus, the total number of unknowns is $2n_b + (n_L + 2n_p - 2)$.

In the implicit singularity programming, the non-singular and singular solutions, $\tilde{\Omega}^{ns}$ and Ω^s (or Ω_f in this case), are simultaneously solved. For $2n_b + (n_L + 2n_p - 2)$

---

[1] Singularity programming is categorized into two schemes: implicit and explicit. The explicit singularity programming is discussed in Sect. 6.1.

unknowns, the same number of equations must be provided. The required equations can be formulated from the boundary conditions along the flow domain and the fracture.

In Formulation III (the dual-variable equivalence), the real and imaginary parts of the nodal equation (Eq. 4.61) are simultaneously used. When the velocity potential P_k is prescribed at the nodal point ζ_k, the approximate boundary value $\tilde{\Phi}(\zeta_k) = \tilde{\Phi}^{\mathrm{ns}}(\zeta_k) + \Phi_{\mathrm{f}}(\zeta_k)$ is presumed to be equal to P_k. As for the stream function, the approximate non-singular boundary value $\tilde{\Psi}^{\mathrm{ns}}(\zeta_k)$ is equated with the non-singular nodal value Ψ_k^{ns}. Then, it follows that

$$
\begin{cases}
\tilde{\Phi}^{\mathrm{ns}}(\zeta_k) + \Phi_{\mathrm{f}}(\zeta_k) = P_k \\
\tilde{\Psi}^{\mathrm{ns}}(\zeta_k) = \Psi_k^{\mathrm{ns}}
\end{cases}
\tag{5.158}
$$

where $\Phi_{\mathrm{f}}(\zeta_k)$ is given by the real part of Eq. 5.155.

In a similar way, when the stream function S_k is prescribed at the nodal point ζ_k, the approximate boundary value $\tilde{\Psi}(\zeta_k) = \tilde{\Psi}^{\mathrm{ns}}(\zeta_k) + \Psi_{\mathrm{f}}(\zeta_k)$ is presumed to be equal to S_k. As for the velocity potential, the approximate non-singular boundary value $\tilde{\Phi}^{\mathrm{ns}}(\zeta_k)$ is equated with the non-singular nodal value Φ_k^{ns}. Then, it follows that

$$
\begin{cases}
\tilde{\Phi}^{\mathrm{ns}}(\zeta_k) = \Phi_k^{\mathrm{ns}} \\
\tilde{\Psi}^{\mathrm{ns}}(\zeta_k) + \Psi_{\mathrm{f}}(\zeta_k) = S_k
\end{cases}
\tag{5.159}
$$

where $\Psi_{\mathrm{f}}(\zeta_k)$ is given by the imaginary part of Eq. 5.155.

These boundary conditions along the flow domain yield a set of $2n_{\mathrm{b}}$ equations for n_{b} nodal points. Thus, the remaining $n_{\mathrm{L}} + 2n_{\mathrm{p}} - 2$ equations need to be obtained from the boundary conditions along the fracture. The condition in terms of the velocity potential, Eq. 5.150, is unconditionally satisfied with $\tilde{\Omega}^{\mathrm{ns}}$, since the non-singular solution gives an identical value of the velocity potential on either side of the fracture. In addition, the fracture singular solution Ω_{f}, Eq. 5.155, is derived so that it satisfies the condition Eq. 5.150. Therefore, the complete solution $\tilde{\Omega} = \tilde{\Omega}^{\mathrm{ns}} + \Omega_{\mathrm{f}}$ unconditionally satisfies Eq. 5.150.

The condition in terms of the stream function, Eq. 5.149, is rewritten as

$$
\Psi_{\mathrm{f}}(z^+) - \Psi_{\mathrm{f}}(z^-) - \frac{k_{\mathrm{f}} w_{\mathrm{f}}}{k}\left(\frac{d\tilde{\Phi}^{\mathrm{ns}}}{ds}(z) + \frac{d\Phi_{\mathrm{f}}}{ds}(z)\right) = 0
\tag{5.160}
$$

where the fact that the non-singular solution gives an identical Ψ on either side of the fracture, $\tilde{\Psi}^{\mathrm{ns}}(z^+) = \tilde{\Psi}^{\mathrm{ns}}(z^-)$, is used. The required $n_{\mathrm{L}} + 2n_{\mathrm{p}} - 2$ equations can be obtained by applying Eq. 5.160 to $n_{\mathrm{L}} + 2n_{\mathrm{p}} - 2$ points along the fracture.

5.7.3.2 Formulation

The fracture singular solution is implicitly coupled with the CVBEM. The approximate boundary value $\tilde{\Omega}(\zeta_k)$ at the nodal point ζ_k is decomposed into the non-singular and fracture singular solutions

$$
\begin{aligned}
2\pi i \tilde{\Omega}(\zeta_k) &= 2\pi i \tilde{\Omega}^{ns}(\zeta_k) + 2\pi i \Omega_f(\zeta_k) \\
&= 2\pi i \tilde{\Omega}^{ns}(\zeta_k) + \sum_{n=1}^{n_L} B_{1n} a_{-n} + \sum_{j=1}^{n_p} B_{2j} b_j + \sum_{j=2}^{n_p-1} B_{3j} c_j \quad (5.161)
\end{aligned}
$$

where $2\pi i \tilde{\Omega}^{ns}(\zeta_k)$ is given by the nodal equation Eq. 4.61 (with Ω_j replaced by Ω_j^{ns}). In the dual-variable equivalence, the boundary conditions Eq. 5.158 and/or Eq. 5.159 are applied to the real and imaginary parts of Eq. 5.161, which yields a set of $2n_b$ equations for n_b nodal points. The coefficients B_{1n}, B_{2j}, and B_{3j} of the fracture parameters $(a_{-n}, b_j,$ and $c_j)$ are detailed in Appendix D.1.

The boundary condition along the fracture, Eq. 5.160, is arranged in terms of unknown variables, resulting in

$$
\sum_{j=1}^{n_b} (C_{1j} \Phi_j^{ns} + C_{2j} \Psi_j^{ns}) + \sum_{n=1}^{n_L} D_{1n} a_{-n} + \sum_{j=1}^{n_p} D_{2j} b_j + \sum_{j=2}^{n_p-1} D_{3j} c_j = 0 \quad (5.162)
$$

which is applied to $n_L + 2n_p - 2$ selected points along the fracture and yields a set of $n_L + 2n_p - 2$ equations. The coefficients C_{1j} and C_{2j} of the non-singular nodal values (Φ_j^{ns} and Ψ_j^{ns}) and the coefficients D_{1n}, D_{2j}, and D_{3j} of the fracture parameters $(a_{-n}, b_j,$ and $c_j)$ are detailed in Appendix D.2.

These equations can be written in matrix form as

$$
\begin{pmatrix} A & B \\ C & D \end{pmatrix} \begin{pmatrix} \Omega^{ns} \\ X \end{pmatrix} = \begin{pmatrix} Y \\ 0 \end{pmatrix} \quad (5.163)
$$

The coefficient matrix is a square matrix of order $2n_b + (n_L + 2n_p - 2)$, which includes the following submatrices:

A: $2n_b \times 2n_b$ coefficient matrix in the dual-variable equivalence
B: $2n_b \times (n_L + 2n_p - 2)$ coefficient matrix of the fracture to the boundary nodes
C: $(n_L + 2n_p - 2) \times 2n_b$ coefficient matrix of the boundary nodes to the fracture
D: $(n_L + 2n_p - 2) \times (n_L + 2n_p - 2)$ coefficient matrix of the fracture to itself

The unknown column vector includes $2n_b$ nodal values:

$$
\Omega^{ns} = (\Phi_1^{ns}, \Psi_1^{ns}, \Phi_2^{ns}, \Psi_2^{ns}, \ldots, \Phi_{n_b}^{ns}, \Psi_{n_b}^{ns})^T \quad (5.164)
$$

and $n_L + 2n_p - 2$ fracture parameters:

$$
X = (a_{-1}, a_{-2}, \ldots, a_{-n_L}, b_1, b_2, \ldots, b_{n_p}, c_2, \ldots, c_{n_p-1})^T \quad (5.165)
$$

The known column vector includes $2n_b$-dimensional subvector $\mathbf{Y}$ which contains $(2\pi P_k, 0)^T$ if P_k is given or $(0, -2\pi S_k)^T$ if S_k is given and $(n_L + 2n_p - 2)$-dimensional zero vector $\mathbf{0}$.

• Solution to Task 6-1

For simplicity, flow velocities at the inlet and outlet boundaries are assumed not to be disturbed by the artificial fracture. Then, the Neumann boundary conditions are the same as those for Task 5-1 (Fig. 4.30).

In the CVBEM, the boundary is modeled with 80 evenly spaced boundary elements of length 0.05 and the fracture of $L_f = 0.5$ and $k_f w_f = 100k$ is modeled with $n_L = 5$ and $n_p = 5$. Figure 5.18 shows the contour plots of $\tilde{\Phi}(z)$ and $\tilde{\Psi}(z)$ computed by the CVBEM, which are the equipotential lines and streamlines, respectively.

A conspicuous impact of the fracture on flow profiles is observed. The equipotential lines detour the fracture, indicating that the velocity potential along the fracture is approximately constant, and the streamlines are approximately perpendicular to the fracture. There are 15 (out of 49) streamlines entering and leaving the fracture. It should be noted that the right edge has inward (flow medium to fracture) flux on both (lower and upper) sides of the fracture, while the left edge has outward flux on both sides. Such a detailed observation would not be possible without accurate modeling of the fracture.

The stream function at $(0.5, 0.5)$ is obtained as $\tilde{\Psi}(0.5 + 0.5i) = 0.651$ by the CVBEM. Since $\Psi(0.5i) = 0$ as defined for the boundary condition, the discharge in the dry area is evaluated as $\Delta q / h = 0.651$. When there is no fracture in the domain, $\Delta q / h = 0.5$ as obtained in Task 5-1; thus, around 30 % increment of $\Delta q / h$ is expected by the installation of the artificial fracture of $L_f = 0.5$ and $k_f w_f = 100k$.

(a)   **(b)**

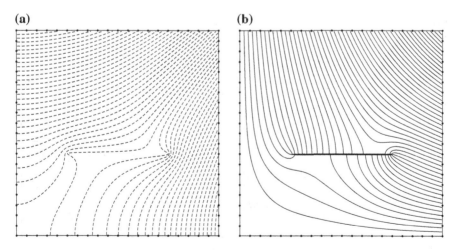

Fig. 5.18 Flow around right-angle faults with an artificial fracture of length $L_f = 0.5$ and conductivity $k_f w_f = 100k$. **a** Equipotential lines. **b** Streamlines

• Solution to Task 6-2

For different values of L_f with a fixed fracture conductivity $k_f w_f = 100k$, the discharge $\Delta q/h$ is evaluated by the CVBEM with the same numerical settings ($n_b = 80$, $n_L = 5$, and $n_p = 5$). Figure 5.19 shows the equipotential lines and streamlines for the case with the fracture of $L_f = 0.8$ and $k_f w_f = 100k$.

As is the case with the fracture of $L_f = 0.5$ (Fig. 5.18), the equipotential lines detour the fracture and the streamlines are approximately perpendicular to the fracture. The fracture behaves as an equipotential object. It is observed that the number of streamlines that pass through the fracture increases for the present case with a longer fracture of $L_f = 0.8$; there are 22 (out of 49) streamlines entering the fracture, flowing along the fracture in the negative x direction, and then leaving the fracture toward the dry area.

Figure 5.20 shows the relation between $\Delta q/h$ and L_f. For longer fractures, higher values of $\Delta q/h$ are expected. When $L_f = 0.8$, for instance, the discharge $\Delta q/h$ becomes 0.736, which is around 47 % higher than that for the case without a fracture. For L_f around 0 and 1, however, the improvement in $\Delta q/h$ with respect to L_f becomes subtle.

The fracture length affects the discharge in the dry area; the longer the length L_f, the higher the discharge $\Delta q/h$. However, the relation is not linear; a too-short or too-long fracture yields little improvement in $\Delta q/h$.

• Solution to Task 6-3

For different values of $k_f w_f/k$ with a fixed fracture length $L_f = 0.5$, the discharge $\Delta q/h$ is evaluated by the CVBEM with the same numerical settings ($n_b = 80$, $n_L = 5$, and $n_p = 5$). Figure 5.21 shows the equipotential lines and streamlines for the case with the fracture of $L_f = 0.5$ and $k_f w_f = 0.1k$.

(a)   **(b)**

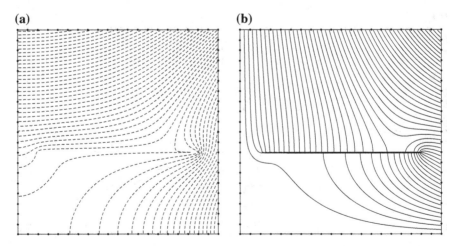

Fig. 5.19 Flow around right-angle faults with an artificial fracture of length $L_f = 0.8$ and conductivity $k_f w_f = 100k$. **a** Equipotential lines. **b** Streamlines

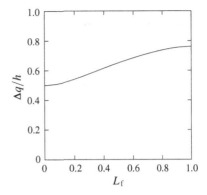

Fig. 5.20 Relation between the discharge in the dry area $\Delta q / h$ and the fracture length L_f

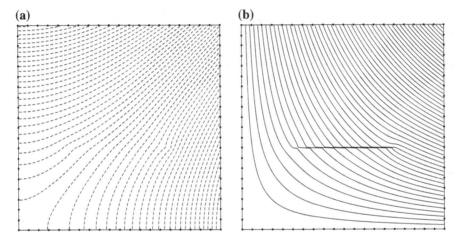

(a)

(b)

Fig. 5.21 Flow around right-angle faults with an artificial fracture of length $L_f = 0.5$ and conductivity $k_f w_f = 0.1k$. **a** Equipotential lines. **b** Streamlines

Compared with the results for the fracture of $k_f w_f = 100k$ (Fig. 5.18), the influence of the fracture on flow profiles becomes less conspicuous, resulting in a smaller improvement in the discharge $\Delta q / h$. The equipotential lines exhibit only a small amount of distortion near the edges of the fracture, and the fracture is no more an equipotential object. Consequently, the streamlines are not perpendicular to the fracture.

Figure 5.22 shows the relation between $\Delta q / h$ and $k_f w_f / k$. For lower values of $k_f w_f$, the discharge $\Delta q / h$ is not much improved; when $k_f w_f / k = 0.1$, $\Delta q / h$ is 0.530, which is only 6 % higher than that for the case without a fracture. For higher values of $k_f w_f$, higher values of $\Delta q / h$ are expected, which is true in the range of $0.1k < k_f w_f < 10k$; $\Delta q / h$ of 0.530 increases to 0.646 as $k_f w_f / k$ increases from 0.1 to 10. However,

Fig. 5.22 Relation between
the discharge in the dry area
$\Delta q / h$ and the fracture
conductivity $k_f w_f$

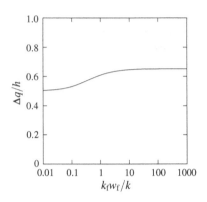

above around $10k$, the fracture conductivity $k_f w_f$ has little influence on $\Delta q / h$; $\Delta q / h$
are 0.646, 0.651, and 0.651 for $k_f w_f / k$ of 10, 100, and 1,000 respectively.

The fracture conductivity affects the discharge in the dry area; the higher the conductivity $k_f w_f$, the higher the discharge $\Delta q / h$. However, the relation is not linear and excessive augmentation in $k_f w_f$ is not necessary.

5.7.4 Flow Over a Thin Shield

Let us consider a thin shield with no conductivity. There is no flow across the impermeable shield, and the boundary condition in terms of the stream function is

$$\Psi = \text{constant} \tag{5.166}$$

along the shield.

The shield singular solution Ω_s has the character opposite to that of the fracture singular solution; Ω_s can be defined such that its imaginary part Ψ_s is continuous and its real part Φ_s is discontinuous across the shield. In the χ plane, it follows that

$$\begin{cases} \Phi_s(\chi^+) - \Phi_s(\chi^-) = \text{Re}[\Omega_s(\chi^+) - \Omega_s(\overline{\chi^+})] \neq 0 \\ \Psi_s(\chi^+) - \Psi_s(\chi^-) = \text{Im}[\Omega_s(\chi^+) - \Omega_s(\overline{\chi^+})] = 0 \end{cases} \tag{5.167}$$

which can be met if the function Ω_s is chosen such that

$$\Omega_s(\chi) = -\overline{\Omega_s(\overline{\chi})} \tag{5.168}$$

In addition, as is the case with the fracture singular solution, Ω_s must be analytic outside the shield, and at points far from the shield, $\chi \to \infty$, it must behave as

$$\lim_{\chi \to \infty} \Omega_s(\chi) = 0 \tag{5.169}$$

The shield singular solution for Eq. 5.168 and 5.169 can be obtained by dividing the fracture singular solution Eq. 5.155 by an imaginary unit i, that is,

$$\Omega_s(\chi) = \frac{1}{i} \sum_{n=1}^{n_L} \frac{a_{-n}}{\chi^n} + \frac{1}{i} \sum_{j=1}^{n_p} \left(\frac{\beta_j \chi_{pj}}{\chi - \chi_{pj}} + \frac{\overline{\beta_j} \chi_{pj}}{\chi - \overline{\chi_{pj}}} \right) \tag{5.170}$$

where $\beta_j = b_j + ic_j$ and a_{-n}, b_j, and c_j are real numbers.

The implicit singularity programming discussed in Sect. 5.7.3 can be applied to a shield. The only differences are the singular solution Eq. 5.170 instead of Eq. 5.155 and the boundary condition Eq. 5.166 instead of Eq. 5.149 along the object. For $n_L + 2n_p - 2$ shield parameters, the same number of equations are obtained by applying the boundary condition

$$\Psi(z_k) - \Psi(z_{k+1}) = 0 \tag{5.171}$$

to $n_L + 2n_p - 2$ selected points along the shield, where $k = 1, 2, \ldots, n_L + 2n_p - 2$ and $z_{n_L+2n_p-1} = z_1$.

Example 5.49 Suppose that the thin object in Motivating Problem 6 is an impermeable shield rather than a conductive fracture. With the same numerical settings ($n_b = 80$, $n_L = 5$, and $n_p = 5$), the CVBEM yields the equipotential lines and streamlines, as shown in Fig. 5.23.

The equipotential lines are perpendicular to the shield and the streamlines detour the object. The stream function at $(0.5, 0.5)$ is obtained as $\tilde{\Psi}(0.5 + 0.5i) = 0.5$ by the CVBEM and the discharge in the dry area is evaluated as $\Delta q / h = 0.5$; there is no change in $\Delta q / h$.

(a)   **(b)**

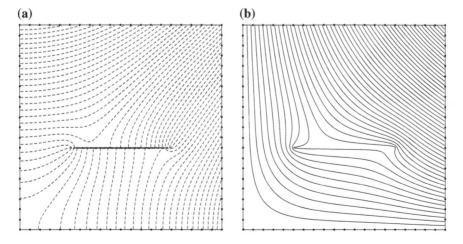

Fig. 5.23 Flow around right-angle faults with a thin shield. **a** Equipotential lines. **b** Streamlines

Chapter 6
Further Applications in Practical Engineering

Motivating Problem 7: Pumping and Injection Wells for Irrigation

Groundwater flows around right-angle impermeable faults with a flow velocity $q_u/A = 2$, as shown in Fig. 6.1. For the purpose of irrigation in the dry area, installation of a pair of wells is planned. Groundwater is withdrawn from the pumping well located at $(0.8, 0.8)$, which is transported through a pipeline to the injection well located at $(0.2, 0.2)$ and returned to the underground for irrigation without land subsidence. Pumping and injection rates per unit thickness are equal in magnitude, being $|q_w/h|$.

Task 7-1 For $|q_w/h| = 1.0$ and $|q_w/h| = 2.0$, draw the groundwater flow profiles around the pumping and injection wells.

Task 7-2 Evaluate the relation between the discharge in the dry area $\Delta q/h$, represented by the amount flowing between $(0, 0.5)$ and $(0.5, 0.5)$, and the flow rate per unit thickness $|q_w/h|$.

Task 7-3 Evaluate the maximum flow rate $|q_w/h|$ that causes no water cycling between the pumping and injection wells, that is, no injected water is withdrawn from the pumping well.

• Solution Strategy to Motivating Problem 7

In the previous chapter, the singularity programing is developed to solve the boundary value problem including the fracture singularity. A similar technique can be applied to the current problem that includes the source/sink singularity.

The source/sink singularity is given by the complex logarithmic function, which is infinitely multi-valued. To ensure the uniqueness of the function, a branch cut is introduce, and thus, care must be taken for a proper configuration of the branch cut.

© Springer International Publishing Switzerland 2015
K. Sato, *Complex Analysis for Practical Engineering*,
DOI 10.1007/978-3-319-13063-7_6

Fig. 6.1 Groundwater
flowing through a dry area
near right-angle
impermeable faults and a
pair of wells

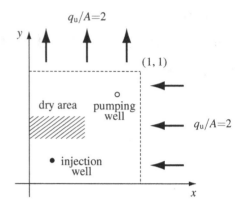

6.1 Explicit Singularity Programming

As is the case with its implicit counterpart, the explicit singularity programming
rests on the principle of superposition. A complete solution Ω is decomposed into
non-singular and singular solutions, Ω^{ns} and Ω^{s}, that is,

$$\Omega = \Omega^{\mathrm{ns}} + \Omega^{\mathrm{s}} \tag{6.1}$$

When the singular solution Ω^{s} is well defined without any unknown parameters, the
singularity programming is performed in an explicit manner.

6.1.1 Singular Solution of Sources and Sinks

The singular solution of a source located at $z_{\mathrm{w}j}$ with a flow rate $q_{\mathrm{w}j}$ (>0) is given by
Eq. 2.111 as

$$\Omega_{\mathrm{w}j}(z) = -\frac{q_{\mathrm{w}j}}{2\pi h} \ln(z - z_{\mathrm{w}j}) \tag{6.2}$$

where h is the thickness of the flow medium. For flow caused by a sink, the same
equation can be used with a negative value of $q_{\mathrm{w}j}$. Note that the parameters $q_{\mathrm{w}j}/h$
and $z_{\mathrm{w}j}$ are usually prescribed and $\Omega_{\mathrm{w}j}(z)$ is well defined.

When multiple sources and/or sinks are involved in the problem, the singular
solution Ω^{s} is obtained by use of superposition. For an n_{w}-source/sink system, the
singular solution becomes

$$\Omega^{\mathrm{s}}(z) = \sum_{j=1}^{n_{\mathrm{w}}} \Omega_{\mathrm{w}j}(z) = -\sum_{j=1}^{n_{\mathrm{w}}} \frac{q_{\mathrm{w}j}}{2\pi h} \ln(z - z_{\mathrm{w}j}) \tag{6.3}$$

which has n_{w} singular points at $z_{\mathrm{w}j}$.

Taking the real part of Eq. 6.3 gives the velocity potential

$$\Phi^{\mathrm{s}}(z) = \sum_{j=1}^{n_{\mathrm{w}}} \mathrm{Re}[\Omega_{\mathrm{w}j}(z)] = -\sum_{j=1}^{n_{\mathrm{w}}} \frac{q_{\mathrm{w}j}}{2\pi h} \ln|z - z_{\mathrm{w}j}| \tag{6.4}$$

and taking the imaginary part gives the stream function

$$\Psi^s(z) = \sum_{j=1}^{n_w} \text{Im}[\Omega_{wj}(z)] = -\sum_{j=1}^{n_w} \frac{q_{wj}}{2\pi h}\theta(z, z_{wj}) \tag{6.5}$$

where $\theta(z, z_{wj})$ is the angle between the vector $z - z_{wj}$ and the branch cut, which can be set arbitrarily for the individual branch point z_{wj}.

6.1.2 Procedure and Formulation

The explicit singularity programming follows a sequence of desuperposition and superposition of singularities. First, the solution and the boundary conditions are decomposed into non-singular and singular components. When the singular solution Ω^s is well defined, Ω^s can be excluded from the CVBEM, which is applied to solve for Ω^{ns}. Finally, the complete solution Ω is recovered by use of superposition: $\Omega = \Omega^{ns} + \Omega^s$.

6.1.2.1 Explicit Scheme

In a boundary value problem containing an n_w-source/sink system, the singular component Ω^s is because of $\sum_{j=1}^{n_w} \Omega_{wj}$, and the approximate solution becomes

$$\tilde{\Omega}(z) = \tilde{\Omega}^{ns}(z) + \sum_{j=1}^{n_w} \Omega_{wj}(z) \tag{6.6}$$

where the non-singular solution $\tilde{\Omega}^{ns}(z)$ is given by Eq. 4.56 and the source/sink singular solution $\Omega_{wj}(z)$ by Eq. 6.2. In the explicit singularity programming, the non-singular solution $\tilde{\Omega}^{ns}$ is solved independently of the singular solution Ω^s. In such a scheme, the boundary conditions in terms of non-singular solutions must be handled in an explicit manner. By use of desuperposition, the non-singular boundary values are decomposed from the prescribed boundary values.

When the velocity potential P_k is prescribed at the nodal point ζ_k, the approximate non-singular boundary value $\tilde{\Phi}^{ns}(\zeta_k)$ is presumed to be equal to $P_k - \Phi_k^s$, where Φ_k^s is the singular component of the velocity potential at ζ_k, given by Eq. 6.4 for $z = \zeta_k$. As for the stream function, the approximate non-singular boundary value $\tilde{\Psi}^{ns}(\zeta_k)$ is equated with the non-singular nodal value Ψ_k^{ns}. Hence, the boundary conditions in terms of non-singular solutions become

$$\begin{cases} \tilde{\Phi}^{ns}(\zeta_k) = P_k - \Phi_k^s \\ \tilde{\Psi}^{ns}(\zeta_k) = \Psi_k^{ns} \end{cases} \tag{6.7}$$

In a similar way, when the stream function S_k is prescribed at the nodal point ζ_k, the approximate non-singular boundary value $\tilde{\Psi}^{ns}(\zeta_k)$ is presumed to be equal to $S_k - \Psi_k^s$, where Ψ_k^s is the singular component of the stream function at ζ_k, given by Eq. 6.5 for $z = \zeta_k$. As for the velocity potential, the approximate non-singular boundary value $\tilde{\Phi}^{ns}(\zeta_k)$ is equated with the non-singular nodal value Φ_k^{ns}. The boundary conditions in terms of non-singular solutions become

$$
\begin{cases}
\tilde{\Phi}^{ns}(\zeta_k) = \Phi_k^{ns} \\
\tilde{\Psi}^{ns}(\zeta_k) = S_k - \Psi_k^s
\end{cases}
\tag{6.8}
$$

The non-singular nodal values Φ_k^{ns} and Ψ_k^{ns} are left unknown at ζ_k, resulting in $2n_b$ unknown boundary values in total. The required $2n_b$ equations can be obtained from the nodal equations in the dual-variable equivalence.

6.1.2.2 Formulation

The approximate non-singular boundary value $\tilde{\Omega}^{ns}(\zeta_k)$ at the nodal point ζ_k is given by the nodal equation, Eq. 4.61 (with Ω_j replaced by Ω_j^{ns}). In the dual-variable equivalence, the boundary conditions Eq. 6.7 and/or Eq. 6.8 are applied to the real and imaginary parts of $\tilde{\Omega}^{ns}(\zeta_k)$, which yields a set of $2n_b$ equations for n_b nodal points.

These equations can be written in matrix form as

$$
\mathbf{A}\mathbf{\Omega}^{ns} = \mathbf{Y}^{ns}
\tag{6.9}
$$

$\mathbf{A}$ is the $2n_b \times 2n_b$ coefficient matrix in the dual-variable equivalence. $\mathbf{\Omega}^{ns}$ is the $2n_b$-dimensional vector which contains unknown non-singular nodal values:

$$
\mathbf{\Omega}^{ns} = (\Phi_1^{ns}, \Psi_1^{ns}, \Phi_2^{ns}, \Psi_2^{ns}, \ldots, \Phi_{n_b}^{ns}, \Psi_{n_b}^{ns})^T
\tag{6.10}
$$

$\mathbf{Y}^{ns}$ is the $2n_b$-dimensional vector that contains $(2\pi(P_k - \Phi_k^s), 0)^T$ if P_k is given or $(0, -2\pi(S_k - \Psi_k^s))^T$ if S_k is given.

6.1.3 Branch Cuts Across a Boundary

As discussed in Sect. 2.2.2, the logarithmic function is infinitely multi-valued, and to keep the function single-valued, a branch cut must be introduced. When source/sink singularities are located interior to the boundary, the branch cuts always intersect the boundary and the stream function caused by the singularities yields discontinuous profiles at the intersections. Such discontinuities are not consistent with the nature of the non-singular solution, which is analytic and must be smooth. In this section, proper configurations of branch cuts are discussed by using example problems.

6.1.3.1 Mixed Boundary Value Problems

Let us consider a source with $q_w/h = 1$ located in the center $z_w = (0.5, 0.5)$ of a square (1×1) flow domain ABCD (Fig. 6.2). The boundary ABCD is impermeable $(\partial \Phi / \partial n|_{ABCD} = 0)$ and the boundary DA is maintained at $\Phi|_{DA} = 0$. The boundary conditions are the Dirichlet type on DA and the Neumann type on ABCD. According to Eq. 4.8, the impermeable boundary ABCD can be modeled by setting a constant value of the stream function. In this problem, $\Psi(\zeta_k) = S_k = 0$ is specified along the boundary ABCD.

In the CVBEM, the boundary is modeled with 80 evenly spaced elements of length 0.05, and two branch-cut configurations are examined: a branch cut intersecting the Dirichlet-type boundary DA (Fig. 6.2a) and a branch cut intersecting the Neumann-type boundary BC (Fig. 6.2b). In the explicit singularity programming, the boundary conditions in terms of non-singular solutions are first evaluated. Along the Dirichlet-type boundary, the non-singular boundary value is given by $P_k - \Phi_k^s$ (Eq. 6.7), and there is no difficulty with the branch cut.

Along the Neumann-type boundary ABCD, on the other hand, the non-singular boundary value is given by $S_k - \Psi_k^s$ (Eq. 6.8), where Ψ_k^s is the singular component of the stream function at ζ_k, defined by

$$\Psi_k^s = -\frac{q_w}{2\pi h} \theta(\zeta_k, z_w) \tag{6.11}$$

The angle $\theta(\zeta_k, z_w)$ is directed from the branch cut to the vector $\zeta_k - z_w$, and thus depends on the configuration of the branch cut. Figure 6.3 shows the profiles of $S_k - \Psi_k^s$ along the boundary ABCD.

When the branch cut intersects the Neumann-type boundary BC, as seen in Fig. 6.3 (shown by open squares), $S_k - \Psi_k^s$ is not continuous but exhibits an abrupt jump at the intersecting point (the middle point of the boundary BC). Since the angle $\theta(\zeta_k, z_w)$ is discontinuous by the amount -2π along the branch cut, Ψ_k^s jumps by the amount $-(q_w/2\pi h)\theta(\zeta_k, z_w) = q_w/h$ across the branch cut, and consequently, $S_k - \Psi_k^s$ jumps by $-q_w/h = -1$ at the intersecting point, as seen in Fig. 6.3.

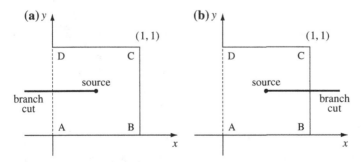

Fig. 6.2 Square flow domain with a mixed boundary condition and two branch-cut configurations. **a** Branch cut intersecting the Dirichlet-type boundary. **b** Branch cut intersecting the Neumann-type boundary

Fig. 6.3 Non-singular
boundary values with a
branch cut intersecting the
Dirichlet-type (*closed
squares*) or Neumann-type
boundary (*open squares*)

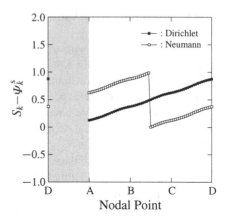

The discontinuous profile of the non-singular boundary values $S_k - \Psi_k^s$ is not consistent with the fact that the non-singular solution is analytic and must be smooth. An improper configuration of the branch cut ruins the CVBEM computation and results in the failure of boundary solutions.

In contrast, when the branch cut intersects the Dirichlet-type boundary DA, $S_k - \Psi_k^s$ is continuous along the Neumann-type boundary ABCD, as seen in Fig. 6.3 (shown by closed squares). This is because of the continuous behavior of Ψ_k^s except along the branch cut. The continuous profile of the non-singular boundary values $S_k - \Psi_k^s$ is consistent with its physical nature, and the CVBEM yields correct non-singular solutions. By use of superposition, the complete solution is properly obtained. Figure 6.4 shows the resultant equipotential lines and streamlines.

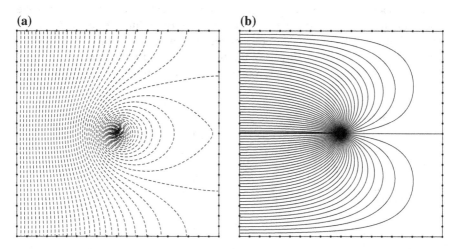

Fig. 6.4 Flow around a source within a mixed Dirichlet/Neumann boundary with a branch cut intersecting the Dirichlet-type boundary. **a** Equipotential lines. **b** Streamlines. (The thick line emanating from the source in the negative x direction corresponds to the branch cut and overlaps a streamline)

The equipotential lines are perpendicular to the Neumann-type boundary ABCD and parallel to the Dirichlet-type boundary DA. Accordingly, the streamlines are parallel to the boundary ABCD and perpendicular to the boundary DA. These flow profiles are consistent with flow physics, which implies that the CVBEM computation works properly with the branch cut intersecting the Dirichlet-type boundary.

For mixed boundary value problems, including source/sink singularities interior to the boundary, to avoid the discontinuity in the non-singular boundary values, the branch cuts from the source(s) and/or sink(s) must be placed so that they intersect the Dirichlet-type boundary.

6.1.3.2 Neumann Boundary Value Problems

Let us consider a multiple-source/sink system in a square (1×1) flow domain ABCD (Fig. 6.5). The boundary DABCD is impermeable $(\partial \Phi / \partial n|_{DABCD} = 0)$ and modeled with the constant boundary value $\Psi(\zeta_k) = S_k = 0$. Note that there is no Dirichlet-type boundary included. A source with $q_{w1}/h = 1$ is located at $z_{w1} = (0.5, 0.2)$ and a sink with $q_{w2}/h = -1$ is located at $z_{w2} = (0.5, 0.8)$, as shown in Fig. 6.5. Since there is no flow across the boundary, the flow rates of source and sink are balanced: $q_{w1}/h + q_{w2}/h = 0$.

In the CVBEM, the boundary is modeled with 80 evenly spaced elements of length 0.05, and two branch-cut configurations are examined: branch cuts focusing on a single point on the boundary (Fig. 6.5a) and branch cuts intersecting the boundary at multiple points (Fig. 6.5b). Along the Neumann-type boundary DABCD, the non-singular boundary value is given by $S_k - \Psi_k^s$ (Eq. 6.8). Figure 6.6 shows the profiles of $S_k - \Psi_k^s$ along the boundary.

When the branch cuts intersect the boundary at multiple points, as seen in Fig. 6.6 (shown by open squares), $S_k - \Psi_k^s$ is not continuous but exhibits abrupt jumps at the intersecting points on the boundary BC. With the same argument in the previous section, $S_k - \Psi_k^s$ jumps by the amount $-q_{w1}/h = -1$ at the intersecting point of the

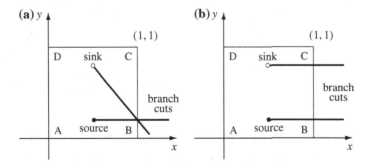

Fig. 6.5 Square flow domain with a Neumann boundary condition and two branch-cut configurations. **a** Branch cuts focusing on a single point on the boundary. **b** Branch cuts intersecting the boundary at multiple points

Fig. 6.6 Non-singular
boundary values with branch
cuts focusing on a single
point (*closed squares*) or
intersecting at multiple
points (*open squares*)

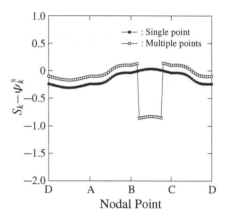

branch cut from the source and by the amount $-q_{w2}/h = 1$ at the intersecting point
of the branch cut from the sink, as seen in Fig. 6.6. These discontinuities result in the
failure of boundary solutions.

It is deduced that the discontinuities caused by several branch cuts can be canceled
out by merging them into a single point. When the branch cuts focus on a single
point on the boundary, $S_k - \Psi_k^s$ is continuous along the boundary DABCD, as seen
in Fig. 6.6 (shown by closed squares). By virtue of the discontinuity cancellation, the
continuity of $S_k - \Psi_k^s$ is maintained, which enables a proper CVBEM computation.
By use of superposition, the complete solution is properly obtained. Figure 6.7 shows
the resultant equipotential lines and streamlines.

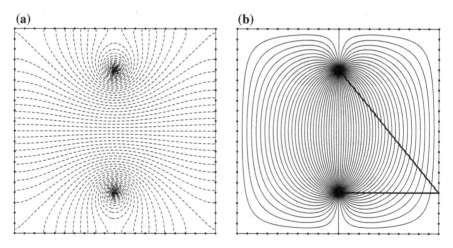

Fig. 6.7 Flow around a source-sink pair within a Neumann-type boundary with branch cuts focusing
on a single point on the boundary. **a** Equipotential lines. **b** Streamlines. (The thick line emanating
from the source and that from the sink towards the focusing point (1, 0.2) correspond to the individual
branch cuts)

The equipotential lines are perpendicular to the Neumann-type boundary, and accordingly, the streamlines are parallel to the boundary, which are consistent with flow physics. It is implied that the CVBEM computation works properly with the branch cuts focusing on a single point on the Neumann-type boundary.

For Neumann boundary value problems, including source/sink singularities interior to the boundary, to avoid the discontinuity in the non-singular boundary values, the branch cuts from the source(s) and sink(s) must be placed so that they focus on a single point on the boundary.

• Solution to Task 7-1

For simplicity, flow velocities at the inlet and outlet boundaries are assumed not to be disturbed by the source or sink. Then, the Neumann boundary conditions are the same as those for Task 5-1 (Fig. 4.30).

In the CVBEM, the boundary is modeled with 80 evenly spaced boundary elements of length 0.05. The branch cut from the injection well located at $(0.2, 0.2)$ and that from the pumping well located at $(0.8, 0.8)$ are placed so that they focus on a single point $(1, 0.2)$ on the boundary. Figure 6.8 shows the resultant equipotential lines and streamlines for $|q_w/h| = 1$.

The streamlines converging towards the pumping well at $(0.8, 0.8)$ are all supplied from the inlet boundary. Consequently, the streamlines emanating from the injection well at $(0.2, 0.2)$ do not flow into the pumping well. There is no underground-flow connection between the injection and pumping wells with $|q_w/h| = 1$. The flow profiles are affected by the value of $|q_w/h|$, as shown in Fig. 6.9 for $|q_w/h| = 2$.

With an increased value of $|q_w/h| = 2$, some of the streamlines emanating from the injection well flow into the pumping well. Hence, there is a water cycling between the pumping and injection wells; a certain amount of injected water is withdrawn

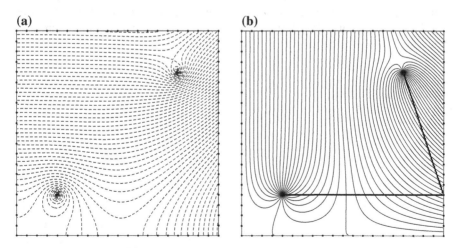

Fig. 6.8 Flow around right-angle faults with pumping and injection wells of $|q_w/h| = 1$. **a** Equipotential lines. **b** Streamlines. (The thick line emanating from the injection well and that from the pumping well towards the focusing point $(1, 0.2)$ correspond to the individual branch cuts)

(a)   **(b)**

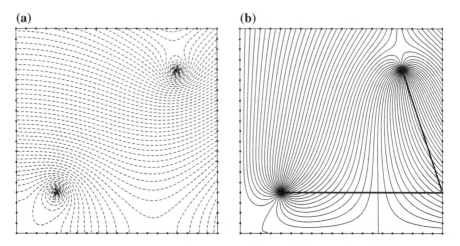

Fig. 6.9 Flow around right-angle faults with pumping and injection wells of $|q_w/h| = 2$. **a** Equipotential lines. **b** Streamlines. (The thick line emanating from the injection well and that from the pumping well towards the focusing point $(1, 0.2)$ correspond to the individual branch cuts)

from the pumping well. A rough estimation of the water cycling can be made by counting the streamlines. Out of 40 streamlines emanating from the injection well, 8 streamlines flow into the pumping well, which implies that 20 % of the injected water is withdrawn from the pumping well.

• Solution to Task 7-2

The discharge in the dry area $\Delta q/h$ is represented by $\Psi(0.5 + 0.5i) - \Psi(0.5i)$ with $\Psi(0.5i) = 0$ as defined for the boundary condition. With the same numerical settings ($n_b = 80$ and the branch cuts focusing on $(1, 0.2)$), the discharge $\Delta q/h = \tilde{\Psi}(0.5 + 0.5i)$ is evaluated by the CVBEM for different values of $|q_w/h|$. Figure 6.10 shows the resultant relation between $\Delta q/h$ and $|q_w/h|$.

Fig. 6.10 Relation between the discharge in the dry area $\Delta q/h$ and the flow rate per unit thickness $|q_w/h|$

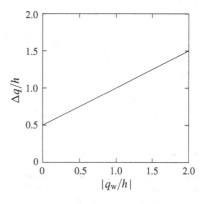

The pumping and injection rate per unit thickness $|q_w/h|$ positively affects the discharge in the dry area $\Delta q/h$; the higher the flow rate $|q_w/h|$, the higher the discharge $\Delta q/h$, which is regressed by the following linear equation:

$$\frac{\Delta q}{h} = 0.5 \left| \frac{q_w}{h} \right| + 0.5 \tag{6.12}$$

For instance, when $|q_w/h| = 1.0$, the discharge $\Delta q/h$ is 1.0, while $|q_w/h| = 2.0$ yields the discharge $\Delta q/h = 1.5$, which is three times as high as the case without a pair of pumping and injection wells ($\Delta q/h = 0.5$ when $|q_w/h| = 0$).

• Solution to Task 7-3

From Task 7-1, it is deduced that the desired flow rate $|q_w/h|$ is between 1 and 2. After a process of trial and error, it is found that $|q_w/h| = 1.5$ gives an approximate solution for the system not to cause a water cycling between the pumping and injection wells. Figure 6.11 shows the equipotential lines and streamlines for $|q_w/h| = 1.5$.

It is confirmed that the streamlines emanating from the injection well do not flow into the pumping well. When $|q_w/h| = 1.5$, the discharge in the dry area is estimated as $\Delta q/h = 1.25$; 150 % increment of $\Delta q/h$ is expected by the installation of the pumping and injection wells without causing a water cycling.

6.1.4 Logarithmic Singularity on a Boundary

The logarithmic function $\ln(z - z_w)$ is not analytic at z_w, which is the singular point. If z_w is on the boundary and located at a nodal point ζ_k, then, there arises a difficulty

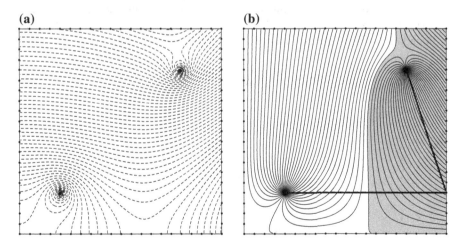

(a)   (b)

Fig. 6.11 Flow around right-angle faults with pumping and injection wells of $|q_w/h| = 1.5$. **a** Equipotential lines. **b** Streamlines. (The thick line emanating from the injection well and that from the pumping well towards the focusing point $(1, 0.2)$ correspond to the individual branch cuts)

caused by the logarithmic singularity since the nodal point ζ_k becomes a singular point. In this section, a proper treatment of the logarithmic singularity is discussed.

6.1.4.1 Five-Spot Pattern

It is common to arrange sources and sinks in regular patterns, thereby forming a specific configuration repeated across the flow domain. For instance, the so-called five-spot pattern consists of a source uniformly surrounded by four sinks, as shown in Fig. 6.12. When the pattern is repeated sufficiently in a large domain, the ratio of sources to sinks approaches 1, and a single element of symmetry becomes representative of all similar five-spot patterns.

Since each quadrant of a five-spot pattern is symmetric, a further geometrical simplification is possible; subdivision of the pattern into four symmetric parts yields a single element of symmetry, ABCD shown in Fig. 6.12. In addition, when the injection and pumping rates are equal in magnitude but opposite in sign, the flow profile also becomes symmetric and a quarter of the discharge from the source to the sink is confined within the flow domain ABCD. There is no flow across the boundary of an individual element, and the boundary DABCD forms a no-flow boundary. A quarter of a five-spot pattern is frequently used as a representative element of the entire flow domain in studying the source-to-sink displacement performances.

6.1.4.2 Boundary Conditions

A square flow domain ABCD (1×1) with a source located at $z_{w1} = (0, 0)$ and a sink at $z_{w2} = (1, 1)$ models a quarter of a five-spot pattern (Fig. 6.13). The injection and pumping rates are $q_{w1}/h = 1$ and $q_{w2}/h = -1$, respectively. Since a quarter of the discharge is confined in the domain, the discharge from the source located at A to the sink at C is 0.25. According to Eq. 2.158, the difference between the values of Ψ along the boundary ABC and the boundary CDA is equal to the discharge from A to C:

$$\Psi_{ABC} - \Psi_{CDA} = \frac{1}{4}\frac{q_{w1}}{h} = 0.25 \qquad (6.13)$$

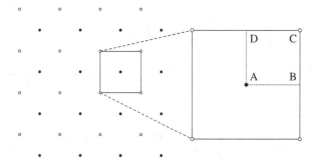

Fig. 6.12 Repeated five-spot pattern; sources (*closed circles*) and sinks (*open circles*)

Fig. 6.13 Neumann
boundary conditions of flow
in a quarter of a five-spot
pattern

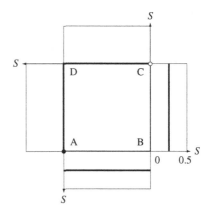

As shown in Fig. 6.13, by setting a constant boundary value $S(\zeta) = 0.25$ along the boundary ABC and a constant boundary value $S(\zeta) = 0$ along the boundary CDA, these boundaries are modeled as no-flow boundaries between which the discharge of 0.25 is specified, as desired.

6.1.4.3 Treatment of Logarithmic Singularities

The logarithmic singularity can be dealt with by considering the actual well geometry; that is, the inside of the well is hollow and must be excluded from the domain of interest. As shown in Fig. 6.14, when z_{wj} coincides with the nodal point ζ_k, ζ_k is expediently shifted to ζ_k': the intersection between the well circumference and the boundary element $\Delta\Gamma_k$. This makes $|\zeta_k' - z_{wj}| = r_{wj}$, where r_{wj} is the wellbore radius, and the singularity problem can be avoided.

A possible problem of discontinuity is caused by the branch cut, as discussed in Sect. 6.1.3. Such a problem can be avoided by placing the branch cut out of the flow domain, as shown in Fig. 6.14, so that it does not intersect the boundary, and thereby the angle in Eq. 6.5 is maintained continuous along the boundary within its principal value $0 \leq \theta(\zeta_k, z_{wj}) < 2\pi$.

Fig. 6.14 Source/sink
located at a nodal point ζ_k,
which is expediently shifted
to ζ_k' to avoid logarithmic
singularity

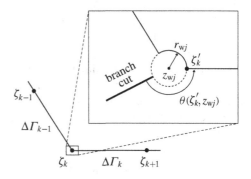

Fig. 6.15 Non-singular
(*closed squares*) and
prescribed (*open squares*)
boundary values for a quarter
of a five-spot pattern

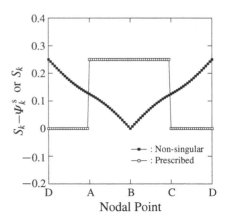

6.1.4.4 Numerical Results[1]

In the CVBEM, the boundary is modeled with 80 evenly spaced elements of length 0.05, to which the Neumann boundary values $S(\zeta)$ are assigned. To ensure the uniqueness of the solution, $\varPhi = 0$ is set at the corner point B. The source and sink radii, r_{w1} and r_{w2}, are both 0.001. Along the Neumann-type boundary DABCD, the non-singular boundary values $S_k - \Psi_k^s$ (Eq. 6.8) are evaluated. Figure 6.15 shows the profiles of $S_k - \Psi_k^s$ and S_k.

The prescribed (complete) boundary value S_k, as seen in Fig. 6.15 (shown by open squares), is not continuous but exhibits abrupt jumps at the source and sink locations (the nodal points A and C) by 0.25 at A and -0.25 at C because of the source/sink singularity. In contrast, $S_k - \Psi_k^s$ is continuous along the Neumann-type boundary DABCD, as seen in Fig. 6.15 (shown by closed squares). For the continuous profile of the non-singular boundary values, the CVBEM yields correct non-singular solutions. By use of superposition, the complete solution is properly obtained. Figure 6.16 shows the approximate boundary values, $\tilde{\varPhi}(\zeta_k)$ and $\tilde{\Psi}(\zeta_k)$, and computational errors, $\tilde{E}_\varPhi(\zeta_k)$ and $\tilde{E}_\Psi(\zeta_k)$.

Since Formulation III equates $\tilde{\Psi}(\zeta_k)$ with S_k, the CVBEM yields no error in $\tilde{\Psi}(\zeta_k)$ along the Neumann-type boundary. Although a tiny amount of error is observed in $\tilde{\varPhi}(\zeta_k)$, as shown in Fig. 6.16a, the singularity programming provides the boundary solutions which are accurate enough for practical purposes. The difference in the velocity potential between the source and the sink is computed as 2.1125. The exact value is known to be 2.1126, and the error is negligible.

Figure 6.17 shows the equipotential lines and streamlines. In the vicinities of the source and sink, the equipotential lines are approximately circular, which implies that the flow near the logarithmic singularity is essentially radial, as is to be expected. The streamlines emanating from the source and converging towards the sink are appropriately simulated.

---

[1] To exemplify the treatment of logarithmic singularities more clearly, the same problem is solved with a smaller number of $n_b = 8$ in Appendix E.2.

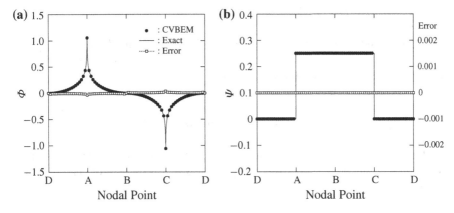

Fig. 6.16 Approximate boundary values, $\tilde{\Phi}(\zeta_k)$ and $\tilde{\Psi}(\zeta_k)$ (*black dots*), exact solutions, $\Phi(\zeta_k)$ and $\Psi(\zeta_k)$ (*solid lines*), and computational errors, $\tilde{E}_\Phi(\zeta_k)$ and $\tilde{E}_\Psi(\zeta_k)$ (*open squares*) for flow in a quarter of a five-spot pattern. **a** Velocity potential. **b** Stream function

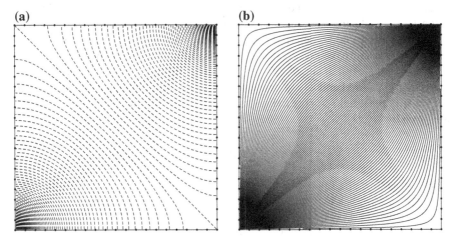

Fig. 6.17 Flow in a quarter of a five-spot pattern. **a** Equipotential lines. **b** Streamlines

These results demonstrate the accuracy of the CVBEM with the singularity programming. By a proper treatment of logarithmic singularities, flow with the source and sink on the boundary can be computed without any difficulty.

6.2 Generalized Singularity Programming

In the previous sections, the implicit and explicit singularity programming schemes are individually developed for different classes of singularities. In the CVBEM, these schemes can be combined and generalized for the problems involving commingled singularities and/or multiple fractures.

6.2.1 Commingled Singularities

The fracture and source/sink singularities can be properly modeled by using the implicit and explicit schemes, respectively. In this section, a generalized scheme is developed for the case when these singularities coexist.

6.2.1.1 Implicit Scheme with Modified Boundary Conditions

In a boundary value problem containing an n_w-source/sink system and a fracture, the singular component Ω^s is from $\sum_{j=1}^{n_w} \Omega_{wj} + \Omega_f$, and the approximate solution becomes

$$\tilde{\Omega}(z) = \tilde{\Omega}^{ns}(z) + \sum_{j=1}^{n_w} \Omega_{wj}(z) + \Omega_f(z) \tag{6.14}$$

where the non-singular solution $\tilde{\Omega}^{ns}(z)$ is given by Eq. 4.56, the source/sink singular solution $\Omega_{wj}(z)$ by Eq. 6.2, and the fracture singular solution $\Omega_f(z)$ by Eq. 5.155. As is the case with the implicit singularity programming, there are $2n_b + (n_L + 2n_p - 2)$ unknown variables to be determined: the nodal values in the non-singular solution $\tilde{\Omega}^{ns}$ and the fracture parameters involved in the fracture singular solution Ω_f.

By using the explicit singularity programming, the source/sink singular solution is explicitly handled and decomposed from the prescribed boundary values. When the velocity potential P_k is prescribed at the nodal point ζ_k, the boundary conditions are given by

$$\begin{cases} \tilde{\Phi}^{ns}(\zeta_k) + \Phi_f(\zeta_k) = P_k - \Phi_k^s \\ \tilde{\Psi}^{ns}(\zeta_k) = \Psi_k^{ns} \end{cases} \tag{6.15}$$

where Φ_k^s is given by Eq. 6.4 for $z = \zeta_k$.

In a similar way, when the stream function S_k is prescribed, the boundary conditions are given by

$$\begin{cases} \tilde{\Phi}^{ns}(\zeta_k) = \Phi_k^{ns} \\ \tilde{\Psi}^{ns}(\zeta_k) + \Psi_f(\zeta_k) = S_k - \Psi_k^s \end{cases} \tag{6.16}$$

where Ψ_k^s is given by Eq. 6.5 for $z = \zeta_k$. For the unknown non-singular nodal values Φ_k^{ns} and Ψ_k^{ns}, these boundary conditions must be satisfied, which yields $2n_b$ equations.

The boundary condition along the fracture, Eq. 5.160, is modified as

$$\Psi_f(z^+) - \Psi_f(z^-) - \frac{k_f w_f}{k}\left(\frac{d\tilde{\Phi}^{ns}}{ds}(z) + \frac{d\Phi_f}{ds}(z)\right) = \frac{k_f w_f}{k} \sum_{j=1}^{n_w} \frac{d\Phi_{wj}}{ds}(z) \tag{6.17}$$

where the source/sink singular solution is explicitly taken into account on the right-hand side. The required $n_L + 2n_p - 2$ equations can be obtained by applying Eq. 6.17 to $n_L + 2n_p - 2$ points along the fracture.

6.2.1.2 Formulation

The fracture singular solution is implicitly coupled with the CVBEM, as discussed in Sect. 5.7.3. The approximate boundary value $\tilde{\Omega}(\zeta_k)$ at the nodal point ζ_k is given by Eq. 5.161. In the dual-variable equivalence, the boundary conditions Eq. 6.15 and/or Eq. 6.16 are applied to the real and imaginary parts of Eq. 5.161, which yields a set of $2n_b$ equations for n_b nodal points.

According to Eq. 6.17, the boundary condition along the fracture, Eq. 5.162, is modified as

$$\sum_{j=1}^{n_b}(C_{1j}\Phi_j^{ns} + C_{2j}\Psi_j^{ns}) + \sum_{n=1}^{n_L}D_{1n}a_{-n} + \sum_{j=1}^{n_p}D_{2j}b_j + \sum_{j=2}^{n_p-1}D_{3j}c_j = F \qquad (6.18)$$

which is applied to $n_L + 2n_p - 2$ selected points along the fracture and yields a set of $n_L + 2n_p - 2$ equations. The constant F corresponds to the source/sink singular solution and is detailed in Appendix D.2.

These equations can be written in matrix form as

$$\begin{pmatrix} A & B \\ C & D \end{pmatrix} \begin{pmatrix} \Omega^{ns} \\ X \end{pmatrix} = \begin{pmatrix} Y^{ns} \\ F \end{pmatrix} \qquad (6.19)$$

The coefficient matrix is a square matrix of order $2n_b + (n_L + 2n_p - 2)$ including the submatrices A, B, C, and D, and the unknown column vector includes $2n_b$ nodal values and $n_L + 2n_p - 2$ fracture parameters, which are the same as those given in Sect. 5.7.3. The known column vector includes $2n_b$-dimensional subvector Y^{ns} which contains $(2\pi(P_k - \Phi_k^s), 0)^T$ if P_k is given or $(0, -2\pi(S_k - \Psi_k^s))^T$ if S_k is given and $(n_L + 2n_p - 2)$-dimensional subvector F which contains the constants F.

Example 6.1 Let us revisit a quarter of a five-spot pattern (1×1) with a source located at $z_{w1} = (0, 0)$ and a sink at $z_{w2} = (1, 1)$, considered in Sect. 6.1.4, and suppose that there exists a fracture of length $L_f = 0.5$ and conductivity $k_f w_f = 10k$ parallel to the x axis with its center at $(0.5, 0.5)$.

In the CVBEM, the boundary is modeled with 80 evenly spaced boundary elements of length 0.05 and the fracture is modeled with $n_L = 5$ and $n_p = 5$. Figure 6.18 shows the equipotential lines and streamlines.

Flow profiles in the vicinities of the source and sink and those around the fracture are appropriately simulated. The generalized singularity programming can handle the commingled singularities.

6.2.2 Multiple Fractures

In a boundary value problem containing an n_w-source/sink system and multiple (n_f) fractures, the singular component Ω^s is from $\sum_{j=1}^{n_w}\Omega_{wj} + \sum_{l=1}^{n_f}\Omega_{fl}$, and the approximate solution becomes

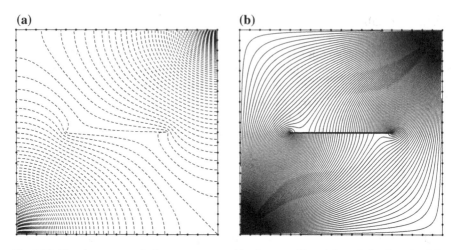

Fig. 6.18 Flow in a quarter of a five-spot pattern with a fracture of length $L_f = 0.5$ and conductivity $k_f w_f = 10k$. **a** Equipotential lines. **b** Streamlines

$$\tilde{\Omega}(z) = \tilde{\Omega}^{ns}(z) + \sum_{j=1}^{n_w} \Omega_{wj}(z) + \sum_{l=1}^{n_f} \Omega_{fl}(z) \tag{6.20}$$

There are $2n_b + n_f(n_L + 2n_p - 2)$ unknown variables to be determined: the nodal values in the non-singular solution $\tilde{\Omega}^{ns}$ and the fracture parameters involved in the fracture singular solution Ω_{fl} for n_f fractures.

The implicit scheme with modified boundary conditions developed in the previous section can be extended, with Ω_f (the single-fracture solution) replaced by $\sum_{l=1}^{n_f} \Omega_{fl}$ (the multiple-fracture solution), to such problems involving multiple fractures. The approximate boundary value $\tilde{\Omega}(\zeta_k)$ at the nodal point ζ_k, Eq. 5.161, is modified as

$$2\pi i \tilde{\Omega}(\zeta_k) = 2\pi i \tilde{\Omega}^{ns}(\zeta_k) + 2\pi i \sum_{l=1}^{n_f} \Omega_{fl}(\zeta_k)$$

$$= 2\pi i \tilde{\Omega}^{ns}(\zeta_k)$$

$$+ \sum_{l=1}^{n_f} \left(\sum_{n=1}^{n_L} B_{1nl} a_{-nl} + \sum_{j=1}^{n_p} B_{2jl} b_{jl} + \sum_{j=2}^{n_p-1} B_{3jl} c_{jl} \right) \tag{6.21}$$

With the modified boundary conditions, the real and imaginary parts of Eq. 6.21 yield a set of $2n_b$ equations for n_b nodal points.

Similarly, the boundary condition along the mth fracture, Eq. 6.18, is modified as

$$\sum_{j=1}^{n_b}(C_{1j}\Phi_j^{ns} + C_{2j}\Psi_j^{ns}) + \sum_{n=1}^{n_L}D_{1n}a_{-nm} + \sum_{j=1}^{n_p}D_{2j}b_{jm} + \sum_{j=2}^{n_p-1}D_{3j}c_{jm}$$

$$+ \sum_{\substack{l=1 \\ l\neq m}}^{n_f}\left(\sum_{n=1}^{n_L}E_{1nl}a_{-nl} + \sum_{j=1}^{n_p}E_{2jl}b_{jl} + \sum_{j=2}^{n_p-1}E_{3jl}c_{jl}\right) = F \qquad (6.22)$$

which is applied to $n_L + 2n_p - 2$ selected points along the mth fracture. For n_f fractures, this yields $n_f(n_L + 2n_p - 2)$ equations in total. The coefficients E_{1nl}, E_{2jl}, and E_{3jl} of the fracture parameters are detailed in Appendix D.2.

These equations can be written in a matrix form as

$$\begin{pmatrix} \mathbf{A} & \mathbf{B}_1 & \mathbf{B}_2 & \mathbf{B}_3 & \cdots & \mathbf{B}_{n_f} \\ \mathbf{C}_1 & \mathbf{D}_1 & \mathbf{E}_{12} & \mathbf{E}_{13} & \cdots & \mathbf{E}_{1n_f} \\ \mathbf{C}_2 & \mathbf{E}_{21} & \mathbf{D}_2 & \mathbf{E}_{23} & \cdots & \mathbf{E}_{2n_f} \\ \mathbf{C}_3 & \mathbf{E}_{31} & \mathbf{E}_{32} & \mathbf{D}_3 & \cdots & \mathbf{E}_{3n_f} \\ \vdots & \vdots & \vdots & \vdots & \ddots & \vdots \\ \mathbf{C}_{n_f} & \mathbf{E}_{n_f1} & \mathbf{E}_{n_f2} & \mathbf{E}_{n_f3} & \cdots & \mathbf{D}_{n_f} \end{pmatrix}\begin{pmatrix} \mathbf{\Omega}^{ns} \\ \mathbf{X}_1 \\ \mathbf{X}_2 \\ \mathbf{X}_3 \\ \vdots \\ \mathbf{X}_{n_f} \end{pmatrix} = \begin{pmatrix} \mathbf{Y}^{ns} \\ \mathbf{F}_1 \\ \mathbf{F}_2 \\ \mathbf{F}_3 \\ \vdots \\ \mathbf{F}_{n_f} \end{pmatrix} \qquad (6.23)$$

The coefficient matrix is a square matrix of order $2n_b + n_f(n_L + 2n_p - 2)$ including the following submatrices:

A: $2n_b \times 2n_b$ coefficient matrix in the dual-variable equivalence

B$_l$: $2n_b \times (n_L + 2n_p - 2)$ coefficient matrix of the lth fracture to the boundary nodes

C$_m$: $(n_L + 2n_p - 2) \times 2n_b$ coefficient matrix of the boundary nodes to the mth fracture

D$_m$: $(n_L + 2n_p - 2) \times (n_L + 2n_p - 2)$ coefficient matrix of the mth fracture to itself

E$_{ml}$: $(n_L + 2n_p - 2) \times (n_L + 2n_p - 2)$ coefficient matrix of the lth fracture to the mth fracture

The unknown column vector includes the nodal values:

$$\mathbf{\Omega}^{ns} = (\Phi_1^{ns}, \Psi_1^{ns}, \Phi_2^{ns}, \Psi_2^{ns}, \dots, \Phi_{n_b}^{ns}, \Psi_{n_b}^{ns})^{T} \qquad (6.24)$$

and the fracture parameters for the mth fracture:

$$\mathbf{X}_m = (a_{-1m}, a_{-2m}, \dots, a_{-n_Lm}, b_{1m}, b_{2m}, \dots, b_{n_pm}, c_{2m}, \dots, c_{n_p-1m})^{T} \qquad (6.25)$$

The known column vector includes $2n_b$-dimensional subvector $\mathbf{Y}^{ns}$ that contains $(2\pi(P_k - \Phi_k^s), 0)^{T}$ if P_k is given or $(0, -2\pi(S_k - \Psi_k^s))^{T}$ if S_k is given and $(n_L + 2n_p - 2)$-dimensional subvectors $\mathbf{F}_m$ that contains the constants F.

Example 6.2 Let us consider a quarter of a five-spot pattern containing multiple fractures. The fracture image is stochastically generated by an unconditional Boolean technique, which randomly distributes 100 fractures in space (parallel to the x axis) with lengths drawn from the normal distribution $N(0.08, 0.03^2)$. The fracture conductivity is $k_f w_f = 10k$.

Except for the fracture configurations, the problem settings and the CVBEM settings are the same as those for Example 6.1. Figure 6.19 shows the equipotential lines and streamlines.

Flow profiles in the presence of distributed multiple fractures are appropriately simulated. By virtue of the semi-analytical nature of the CVBEM, the technique of singularity programming can be applied to treat commingled singularities with multiple fractures analytically, which ensures the computational stability and accuracy of the numerical scheme.

Motivating Problem 8: Interwell Tracer Test

For the purpose of characterizing the flow medium, an interwell tracer test is planned in a quarter of a five-spot pattern. From the source located at $(0, 0)$, a tracer slug is injected for the time interval $\Delta t = 0.2$, and, in turn, is followed by a tracer-free fluid. The effluent tracer-slug concentration C is monitored at the sink located at $(1, 1)$. If the flow medium is homogeneous and hydrodynamic dispersion of the tracer is negligible, the profile of effluent concentration is obtained as shown in Fig. 6.20.

Task 8-1   Reproduce Fig. 6.20 by the CVBEM to confirm the solution.
Task 8-2   Predict the breakthrough time of the tracer and the profile of effluent tracer-slug concentration when there exists a fracture of length $L_f = 0.5$

(a)   **(b)**

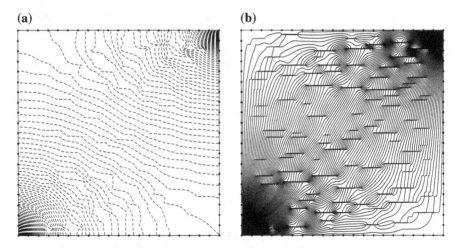

Fig. 6.19 Flow in a quarter of a five-spot pattern with 100 randomly-distributed fractures of conductivity $k_f w_f = 10k$. **a** Equipotential lines. **b** Streamlines

and conductivity $k_f w_f = 10k$ parallel to the x axis with its center at $(0.5, 0.5)$.

Task 8-3 Redo Task 8-2 for the case when hydrodynamic dispersion is not negligible and evaluate the effects of dispersivity on the tracer performances.

• Solution Strategy to Motivating Problem 8

The streamlines from the source to the sink can be drawn by contouring the stream functions, which, however, convey no temporal information. To simulate interwell tracer tests, streamlines need to be tracked (not drawn) with an explicit evaluation of elapsed time.

Another task is to include a different class of flow physics (dispersivity) in the flow computation. Several types of one-dimensional flow problems have been investigated and the corresponding solutions are provided, either analytically or semi-analytically, which are coupled with the CVBEM through the technique of streamline simulation.

6.3 Streamline Tracking

Since streamlines are the level curves of the stream function, it is natural to use the stream function for tracking streamlines, which is covered in this section. In addition, for the sake of completeness, a traditional way of streamline tracking, the velocity vector method, is briefly reviewed.

6.3.1 Velocity Vector Method

Streamlines can be defined as the instantaneous curves that are at every point tangent to the direction of the velocity $\mathbf{V} = \overline{W}$ at that point, where W is the complex velocity

Fig. 6.20 Effluent tracer-slug concentration for a slug size of $\Delta t = 0.2$

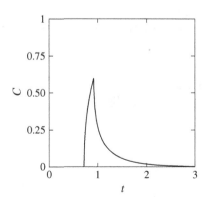

defined by Eq. 2.120. Hence, the unit tangent vector at z is given by

$$\mathbf{u} = \frac{dz}{ds} = \frac{\overline{W(z)}}{|W(z)|} \qquad (6.26)$$

where s is the arc length along the streamline. The streamline tracking by using this relation is called the velocity vector method, which facilitates the Euler or modified Euler method.

6.3.1.1 Euler Method

Let z_j be the current position of an imaginary fluid particle and $z_{j+1/2}$ be the tracked point, as shown in Fig. 6.21. Assuming that the velocity between z_j and $z_{j+1/2}$ is given by $\overline{W_j} = \overline{W(z_j)}$ and integrating Eq. 6.26 from z_j to $z_{j+1/2}$ yields

$$z_{j+1/2} = z_j + \frac{\overline{W_j}}{|W_j|} \Delta s \qquad (6.27)$$

where Δs is the distance between z_j and $z_{j+1/2}$. This approximation is known as the Euler method.

The time increment $\Delta \tau_j$ required for the particle to move from z_j to $z_{j+1/2}$ is obtained as

$$\Delta \tau_j = \frac{\Delta s}{V_{sj}} = \phi \frac{\Delta s}{|W_j|} \qquad (6.28)$$

where the distance Δs is divided by the seepage velocity, $V_{sj} = |W_j|/\phi$, evaluated at z_j.

6.3.1.2 Modified Euler Method

Let z_j be the current position of an imaginary fluid particle and z_{j+1} be the tracked point, as shown in Fig. 6.22. Assuming that the velocity between z_j and z_{j+1} is given

Fig. 6.21 Streamline tracking using the Euler method

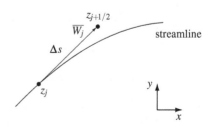

Fig. 6.22 Streamline
tracking using the modified
Euler method

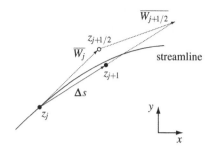

by the average of the velocities at z_j and $z_{j+1/2}$, $(\overline{W_j} + \overline{W_{j+1/2}})/2 = (\overline{W(z_j)} + \overline{W(z_{j+1/2})})/2$ and integrating Eq. 6.26 from z_j to z_{j+1} yields

$$z_{j+1} = z_j + \frac{\overline{W_j} + \overline{W_{j+1/2}}}{|W_j + W_{j+1/2}|} \Delta s \qquad (6.29)$$

where Δs is the distance between z_j and z_{j+1}. This approximation is known as the modified Euler method.

The time increment $\Delta\tau_j$ required for the particle to move from z_j to z_{j+1} is obtained as

$$\Delta\tau_j = \frac{\Delta s}{V_{sj}} = \phi \frac{2\Delta s}{|W_j + W_{j+1/2}|} \qquad (6.30)$$

where the distance Δs is divided by the average seepage velocity over the interval, $V_{sj} = |W_j + W_{j+1/2}|/(2\phi)$.

As is obvious from Figs. 6.21 and 6.22, the accuracy of the velocity vector method depends on Δs. Although smaller values of Δs yield higher accuracy, it is not ensured that the current point and the point tracked by the velocity vector method are exactly on the same streamline.

6.3.2 Stream Function Method

Since streamlines are defined as curves of constant stream function

$$\Psi = \Psi(x, y) = \text{constant} \qquad (6.31)$$

they can be tracked by finding the points of a constant value of Ψ. Let z_j be a point on a certain streamline, which has a stream function of Ψ_0. The Euler method with Δs_0 gives the first-order estimate of the provisional point $z_{j+1/2}$ as shown in Fig. 6.23.

Fig. 6.23 Streamline
tracking using the stream
function method

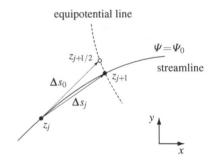

To improve the tracking accuracy, the second approximation can be obtained by the Newton–Raphson procedure along the equipotential line that passes through $z_{j+1/2}$. The complex velocity $W_{j+1/2}$ at $z_{j+1/2}$ can be expressed with the complex potentials at $z_{j+1/2}$ and z_{j+1} as

$$W_{j+1/2} = -\frac{d\Omega}{dz}(z_{j+1/2}) \simeq -\frac{\Omega_{j+1} - \Omega_{j+1/2}}{z_{j+1} - z_{j+1/2}}$$

$$= -\frac{\Phi_{j+1} + i\Psi_{j+1} - \Phi_{j+1/2} - i\Psi_{j+1/2}}{z_{j+1} - z_{j+1/2}} \tag{6.32}$$

Since $z_{j+1/2}$ and z_{j+1} are on the same equipotential line, $\Phi_{j+1/2} = \Phi_{j+1}$. Also, since z_{j+1} and z_j must be on the same streamline, $\Psi_{j+1} = \Psi_0$. Thus, Eq. 6.32 reduces to

$$W_{j+1/2} = -i\frac{\Psi_0 - \Psi_{j+1/2}}{z_{j+1} - z_{j+1/2}} \tag{6.33}$$

Solving Eq. 6.33 for z_{j+1} yields

$$z_{j+1} = z_{j+1/2} + i\frac{\Psi_{j+1/2} - \Psi_0}{W_{j+1/2}} \tag{6.34}$$

Equation 6.34 relates a provisional point $z_{j+1/2}$ and the second approximation z_{j+1}. If the stream function $\Psi_{j+1/2}$ at $z_{j+1/2}$ is equal to Ψ_0, Eq. 6.34 indicates $z_{j+1} = z_{j+1/2}$ and both points are on the streamline that passes through z_j. If $\Psi_{j+1/2} \neq \Psi_0$, Eq. 6.34 is applied recursively to obtain z_{j+1} until $|\Psi_{j+1/2} - \Psi_0|$ becomes sufficiently small. This process is equivalent to the Newton–Raphson method for solving the equation

$$\Psi(z) - \Psi_0 = 0 \tag{6.35}$$

for z, and is referred to as the stream function method. In theory, the stream function method gives the exact loci of streamlines.

The time increment $\Delta\tau_j$ required for the particle to move from z_j to z_{j+1} is obtained as

$$\Delta\tau_j = \frac{\Delta s_j}{V_{sj}} = \phi\frac{2|z_{j+1} - z_j|}{|W_j + W_{j+1}|} \tag{6.36}$$

where the distance $\Delta s_j = |z_{j+1} - z_j|$ is divided by the average seepage velocity over the interval, $V_{sj} = |W_j + W_{j+1}|/(2\phi)$. Note that Δs_j is not necessarily equal to Δs_0 used for the provisional point.

6.3.3 Evaluation of Complex Velocities

It is revealed that the complex velocity W plays a key role in the streamline tracking methods. In the generalized singularity programming for a boundary value problem containing an n_w-source/sink system and n_f fractures, the approximate solution is given by Eq. 6.20, and the corresponding complex velocity is obtained as

$$W = -\frac{d\tilde{\Omega}}{dz}(z) = -\frac{d\tilde{\Omega}^{ns}}{dz}(z) - \sum_{j=1}^{n_w}\frac{d\Omega_{wj}}{dz}(z) - \sum_{l=1}^{n_f}\frac{d\Omega_{fl}}{dz}(z) \tag{6.37}$$

where each term can be evaluated analytically.

The non-singular solution $\tilde{\Omega}^{ns}(z)$ is given by Eq. 4.56, and its derivative with respect to z becomes

$$\frac{d\tilde{\Omega}^{ns}}{dz}(z) = \frac{1}{2\pi i}\sum_{j=1}^{n_b}\frac{\Omega_{j+1}^{ns} - \Omega_j^{ns}}{\zeta_{j+1} - \zeta_j}\ln\frac{\zeta_{j+1} - z}{\zeta_j - z} \tag{6.38}$$

where $\Omega_j^{ns} = \Phi_j^{ns} + i\Psi_j^{ns}$ is the non-singular nodal value at the nodal point ζ_j.

For the singular solution of a source or sink located at z_{wj} with a flow rate q_{wj}, given by Eq. 6.2, its derivative with respect to z becomes

$$\frac{d\Omega_{wj}}{dz}(z) = -\frac{q_{wj}}{2\pi h}\frac{1}{z - z_{wj}} \tag{6.39}$$

where the singularity at $z = z_{wj}$ is handled as discussed in Sect. 6.1.4.

For the fracture singular solution $\Omega_{fl}(z)$, by use of the chain rule, its derivative with respect to z becomes

$$\frac{d\Omega_{fl}}{dz}(z) = \frac{d\chi}{dz}\frac{d\Omega_{fl}}{d\chi}(\chi) \tag{6.40}$$

From Eq. 5.151, for the lth fracture with end points z_{1l} and z_{2l}, $d\chi/dz$ is given by

$$\frac{d\chi}{dz} = \frac{d\chi}{dZ}\frac{dZ}{dz} = \frac{2\chi^2}{\chi^2 - 1}\frac{2}{z_{2l} - z_{1l}} = \frac{4\chi^2}{(\chi^2 - 1)(z_{2l} - z_{1l})} \tag{6.41}$$

where $Z = (\chi + 1/\chi)/2$ is used. From Eq. 5.155, the derivative of $\Omega_{fl}(\chi)$ with respect to χ is given by

$$\frac{d\Omega_{fl}}{d\chi}(\chi) = -\sum_{n=1}^{n_L} \frac{na_{-nl}}{\chi^{n+1}} - \sum_{j=1}^{n_p} \left[\frac{\beta_{jl}\chi_{pj}}{(\chi - \chi_{pj})^2} + \frac{\overline{\beta_{jl}\chi_{pj}}}{(\chi - \overline{\chi_{pj}})^2} \right] \tag{6.42}$$

Substituting Eqs. 6.41 and 6.42 into Eq. 6.40 yields the desired derivative.

6.3.4 Streamlines Through Fractures

When there exists a fracture in the flow domain, some of the streamlines may flow into and out of the fracture, as exemplified in Fig. 6.24a. The streamline with $\Psi = \Psi_0$ flows into the fracture at z_{in} and out of the fracture at z_{out}. Figure 6.24b shows the corresponding velocity vector field.

The velocity vectors are discontinuous around the fracture, as seen in Fig. 6.24b, and do not convey sufficient information for relating a certain inflow point with the corresponding outflow point. Hence, once the streamline flows into the fracture, the corresponding outward streamline cannot be identified by the velocity vector method.

The stream function method overcomes this difficulty. To check whether or not a certain streamline flows into the fracture, the maximum and minimum values of the stream function along the fracture are first computed. If Ψ_0 is between these extrema, the streamline with Ψ_0 flows into the fracture.

Although the inflow point z_{in} can be found in the z plane, the outflow point z_{out} must be identified in the transformed χ plane (Fig. 5.17). This is because the $+$ and $-$

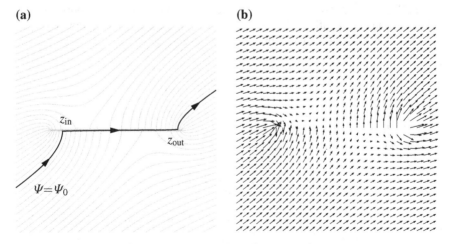

(a)   **(b)**

Fig. 6.24 Uniform inclined flow through a fracture. **a** Streamline with $\Psi = \Psi_0$ flowing into and out of the fracture. **b** Velocity vector field

sides of the fracture (Fig. 5.16) can be distinguished only in the χ plane. The outflow point χ_{out} in the χ plane can be obtained by finding the roots for the equation

$$\Psi(\chi_{out}) - \Psi_0 = 0 \tag{6.43}$$

Then, χ_{out} is back-transformed to z_{out} in the z plane through the transformation Eq. 5.151.

6.3.5 Tracer Transport

A means to follow underground fluid movement is an invaluable tool for evaluating hydrodynamic characteristics. Interwell tracer tests have been conducted for this purpose, in which easy-to-detect materials are injected at a source and monitored at a sink. For the tracer flow behavior, two types of transport mechanisms are considered: convection and convection-dispersion processes.

6.3.5.1 Convection Equation

Along the direction of irrotational flow, the convection equation for incompressible fluids, under appropriate assumptions, is given by

$$\frac{\partial C}{\partial t} + V_s \frac{\partial C}{\partial s} = 0 \tag{6.44}$$

where C is the concentration, s the arc length along the flow direction, and V_s the seepage velocity. When V_s varies along s, the concept of time of flight (TOF) is of use. The TOF τ is the transformed spatial coordinate which is equal to the time required for the fluid particle to move from $s_0 (= 0)$ to s_1, defined by

$$\tau(s_1) = \int_0^{s_1} \frac{ds}{V_s(s)} \tag{6.45}$$

through which the velocity variation can be incorporated implicitly with the resultant differential equation.

The TOF yields the identity

$$\frac{d\tau}{ds} = \frac{1}{V_s} \tag{6.46}$$

with which Eq. 6.44 reduces to

$$\frac{\partial C}{\partial t} + \frac{\partial C}{\partial \tau} = 0 \tag{6.47}$$

The solution of Eq. 6.47 is trivial; any initial configuration shifts to the right (in the direction of τ).

Subject to the initial and boundary conditions as

$$\begin{cases} C(s, t = 0) = 0 & (s \geq 0) \\ C(s = 0, t) = C_0 & (t > 0) \end{cases} \tag{6.48}$$

the solution is given by

$$C(\tau, t) = \begin{cases} 0 & (t < \tau) \\ C_0 & (t > \tau) \end{cases} \tag{6.49}$$

In the context of tracer flow, the solution Eq. 6.49 states that the time required for the tracer front (with the concentration C_0) to reach s_1 is given by $t = \tau(s_1)$.

6.3.5.2 Convection-Dispersion Equation

The convection-dispersion equation for incompressible fluids, under appropriate assumptions, is obtained by adding a dispersion term to the convection equation as

$$\frac{\partial C}{\partial t} + V_s \frac{\partial C}{\partial s} - \alpha_L V_s \frac{\partial^2 C}{\partial s^2} = 0 \tag{6.50}$$

where α_L is the longitudinal dispersivity. Variations of velocity V_s are incorporated implicitly through a set of coordinate transformations τ and ω. The TOF τ is given by Eq. 6.45 and ω is defined by

$$\omega(s_1) = \int_0^{s_1} \frac{ds}{V_s^2(s)} \tag{6.51}$$

By using a boundary layer approximation with τ and ω, Eq. 6.50 reduces to

$$\frac{\partial C}{\partial \omega} - \alpha_L \frac{\partial^2 C}{\partial \eta^2} = 0 \tag{6.52}$$

where $\eta = \tau - t$. Equation 6.52 is a simple diffusion equation, the solutions of which for various initial and boundary conditions are known.

Subject to the initial and boundary conditions as

$$\begin{cases} C(s, t = 0) = 0 & (s \geq 0) \\ C(s = 0, t) = C_0 & (t > 0) \\ C(s \to \infty, t) = 0 & (t > 0) \end{cases} \tag{6.53}$$

the solution of Eq. 6.52 is given by

$$C(\tau, t) = \frac{C_0}{2}\text{erfc}\left(\frac{\tau - t}{\sqrt{4\alpha_L \omega}}\right) \tag{6.54}$$

where erfc is the complementary error function.

6.3.6 Streamline Simulation

The fundamental idea of streamline simulation is to decompose a two-dimensional flow problem into multiple one-dimensional problems to be solved along stream-lines. Along each streamline, one-dimensional flow is analytically modeled, and the complete two-dimensional solution is recovered by combining the individual one-dimensional solutions.

6.3.6.1 Convection Processes

The breakthrough time and the profile of effluent tracer concentration are of main interest in the tracer tests, which can be simulated by the streamline tracking. Let us track n_s streamlines from equiangular points around the wellbore of the source. The time τ_{bi} required for the ith streamline to reach the sink is obtained by approximating Eq. 6.45 along the streamline coordinate s as

$$\tau_{bi} = \sum_{j=1}^{m_i} \frac{\Delta s_j}{V_{sj}} = \sum_{j=1}^{m_i} \Delta\tau_j \tag{6.55}$$

where m_i is the number of trackings from the source to the sink along the ith streamline and $\Delta\tau_j$ is given by Eq. 6.36. After all the streamlines are tracked to the sink, the breakthrough time t_b of the tracer can be estimated from the fastest streamline that arrives at the sink:

$$t_b = \min(\tau_{b1}, \tau_{b2}, \ldots, \tau_{bn_s}) \tag{6.56}$$

The discharge Δq_i of the ith streamline is proportional to the velocity $|W_i|$ at the emanating point, and is given by

$$\Delta q_i = \frac{|W_i|}{\sum_{j=1}^{n_s} |W_j|} q \tag{6.57}$$

where $\sum_{i=1}^{n_s} \Delta q_i = q$. The tracer effluent concentration $C(t)$ at the sink can be constructed numerically by

$$C(t) = \frac{1}{q} \sum_{i=1}^{n_s} C(\tau_{bi}, t)\Delta q_i \tag{6.58}$$

where $C(\tau_{bi}, t)$ is given by Eq. 6.49. Equation 6.58 states that if τ_{bi} of the ith streamline is smaller than t, the concentration assigned to this streamline contributes to the effluent concentration.

6.3.6.2 Convection-Dispersion Processes

Because of the dispersive process, the interface between the tracer front and the native fluid becomes indistinct. In evaluating the coordinate transformations, Eqs. 6.45 and 6.51, s_1 is assumed to be the hypothetical tracer front that would exist if no dispersion occurred. Then, τ_{bi} is given by Eq. 6.55 and the corresponding ω is approximated as

$$\omega_{bi} = \sum_{j=1}^{m_i} \frac{\Delta s_j}{V_{sj}^2} = \sum_{j=1}^{m_i} \frac{\Delta \tau_j^2}{\Delta s_j} \tag{6.59}$$

where m_i is the number of trackings required for the hypothetical front to reach the sink along the ith streamline.

The tracer effluent concentration $C(t)$ at the sink can be constructed by Eq. 6.58, where $C(\tau_{bi}, t)$ is given by Eq. 6.54. Unlike the convection process, the concentration assigned to the ith streamline may contribute to the effluent concentration prior to τ_{bi} because of the dispersed profile of the tracer front.

• Solution to Task 8-1

For the interwell tracer test in a quarter of a five-spot pattern, the same boundary conditions as those in Sect. 6.1.4 apply. In the CVBEM, the boundary is represented by arranging 80 evenly spaced boundary elements. From 201 equiangular points around the wellbore of radius 0.001, streamlines are tracked by the stream function method with $\Delta s_0 = 0.005$.

For the continuous tracer injection with $C_0 = 1$ at the source, the streamline simulation is applied. Hydrodynamic dispersion is negligible; thus, the convection solution Eq. 6.49 is used. Figure 6.25 shows the resultant effluent tracer concentration at the sink, where the stepwise profile is produced by the discrete streamline tracking. Also shown is the analytical solution.[2] The agreement between the CVBEM result and the exact solutions is excellent. In particular, the breakthrough time of $t_b = 0.718$ is accurately predicted.

Figure 6.26a shows the streamlines and tracer fronts at breakthrough, where the tracer fluid is indicated by the black lines and the native fluid by the gray lines. The

---

[2] The effluent concentration $C(t)$ is related to t through the following equations:

$$\begin{cases} t = 0.228473(1 + \lambda)K(\lambda^2) \\ \lambda = \tan^2 \dfrac{\pi(1 - C(t))}{4} \end{cases}$$

where K is the complete elliptic integral of the first kind.

Fig. 6.25 Effluent tracer concentration in a quarter of a five-spot pattern

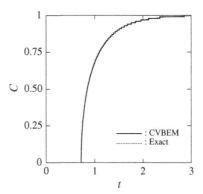

leading interface between the tracer and native fluids is characterized by its cusp shape. This is because the fastest streamline is on the diagonal between the source and sink. The slower streamlines travel near the boundaries.

In the tracer slug test, a tracer fluid of a slug size Δt is injected, and, in turn, is followed by a tracer-free (native) fluid. The trailing interface between the tracer and tracer-free fluids is behind the leading interface by the amount Δt. Such streamline profiles are shown in Fig. 6.26b for $\Delta t = 0.2$. Consequently, the effluent concentration profile of the tracer-free fluid (C_N) is delayed from that of the tracer fluid (C_T) by the amount Δt, as shown in Fig. 6.27a.

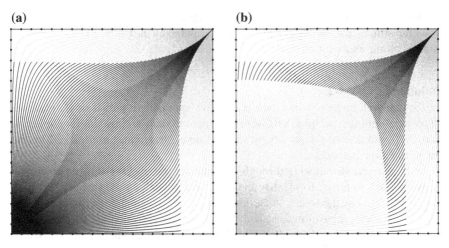

(a) **(b)**

Fig. 6.26 Tracer (*black lines*) and tracer-free (*gray lines*) fluids at breakthrough in a quarter of a five-spot pattern. **a** Continuous tracer injection. **b** Tracer slug injection with $\Delta t = 0.2$. (For visual clarity, only 101 streamlines are drawn, although 201 streamlines are tracked to generate the effluent tracer concentration profile $C(t)$)

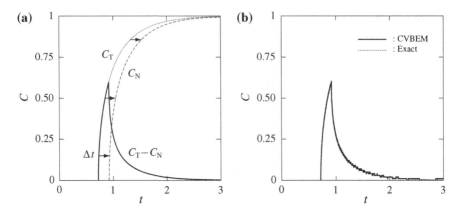

Fig. 6.27 Effluent tracer-slug concentration graphically constructed by using the profiles of continuous tracer injection. **a** Effluent concentrations of the tracer (C_T) and tracer-free (C_N) fluids. **b** Effluent tracer-slug concentrations for a slug size of $\Delta t = 0.2$ simulated by the CVBEM and by the analytical solution

The effluent tracer-slug concentration $C(t)$, therefore, can be obtained by subtracting the concentration profile of the tracer-free fluid $C_N(t)$ from that of the tracer $C_T(t)$. By using the identity $C_N(t) = C_T(t - \Delta t)$, it follows

$$C(t) = C_T(t) - C_T(t - \Delta t) \tag{6.60}$$

which implies that $C(t)$ can be constructed by using only $C_T(t)$ values. Figure 6.27b shows the effluent tracer-slug concentration computed by the CVBEM, which is consistent with the exact solution.

• Solution to Task 8-2

For the interwell tracer test in a quarter of a five-spot pattern with a single fracture, the problem settings and the CVBEM settings are the same as those for Example 6.1. From 201 equiangular points around the wellbore, streamlines are tracked by the stream function method.

For the streamlines flowing through the fracture, the tracking scheme discussed in Sect. 6.3.4 is applied. To identify the outflow points along the fracture, Eq. 6.43 is solved for χ_{out} by the bisection method, which is based on the intermediate value theorem and yields robust solutions without the need for derivatives of Ψ. Figure 6.28 shows the streamlines and the tracer distributions at different times: $t = 0.25, 0.529$ (at breakthrough), 0.75, and 1.05, where the fluid mixing within the fracture is assumed to be negligible.

It is observed that the streamlines can be grouped into two sets: those flowing through the fracture and the others not through the fracture. Owing to the high

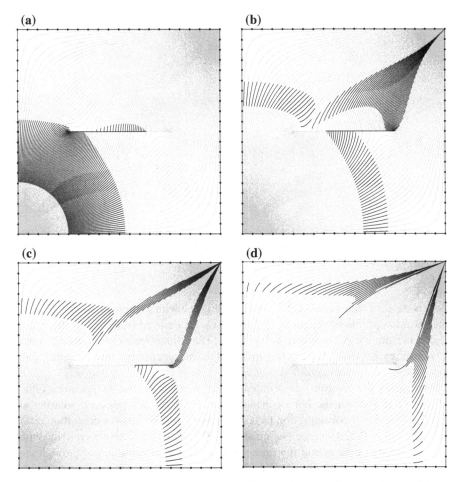

Fig. 6.28 Tracer (*black lines*) and tracer-free (*gray lines*) fluids for a slug size of $\Delta t = 0.2$ in a quarter of a five-spot pattern containing a fracture of length $L_f = 0.5$ and conductivity $k_f w_f = 10k$. **a** $t = 0.25$. **b** $t_b = 0.529$ at breakthrough. **c** $t = 0.75$. **d** $t = 1.05$. (For visual clarity, only 101 streamlines are drawn, although 201 streamlines are tracked to generate the effluent tracer concentration profile $C(t)$)

conductivity of the fracture, the tracer through the fracture is transported faster and thus reaches the sink earlier. The breakthrough occurs at $t_b = 0.529$, earlier than that of the homogeneous case (0.718). The tracer fluid distribution at breakthrough is shown in Fig. 6.28b.

The tracer is also transported through the second group of the streamlines not flowing through the fracture, which are visually confirmed in Fig. 6.28c, d. In particular, Fig. 6.28d shows the tracer distribution when the fastest streamline not through the fracture reaches the sink at $t = 1.05$.

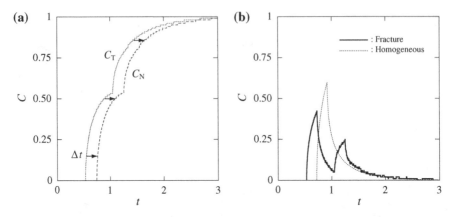

Fig. 6.29 Effluent tracer-slug concentration graphically constructed by using the profiles of continuous tracer injection in the presence of a fracture. **a** Effluent concentrations of the tracer (C_T) and tracer-free (C_N) fluids. **b** Effluent tracer-slug concentration for a slug size of $\Delta t = 0.2$

As is the case with Task 8-1, for the tracer slug test with a slug size Δt, the effluent concentration profile of the tracer-free fluid (C_N) is delayed from that of the tracer (C_T) by the amount Δt, as shown in Fig. 6.29a. The effluent tracer-slug concentration $C(t)$ can be expressed by Eq. 6.60 regardless of the presence of fractures. Figure 6.29b shows the resultant concentration profile.

Because of the fracture, the effluent tracer-slug concentration becomes quite different from that for the homogeneous case. The breakthrough occurs earlier at $t_b = 0.529$ and two concentration peaks are observed. The first concentration peak outbreaks at $t = 0.529$ and the second at $t = 1.05$, which respectively coincide with the breakthrough time t_b and the time when the fastest streamline not through the fracture reaches the sink. This implies that the first concentration peak corresponds to the tracer flowing through the fracture and the second corresponds to the tracer not through the fracture.

• Solution to Task 8-3

Except for the dispersivity, the problem settings and the CVBEM settings are the same as those for Task 8-2. From 201 equiangular points around the wellbore of radius 0.001, streamlines are tracked by the stream function method. When hydrodynamic dispersion is not negligible, the convection-dispersion solution Eq. 6.54 is used in the streamline simulation. Figure 6.30 shows the tracer distributions at $t = 0.5$ for different dispersivities: $\alpha_L = 0$ (no dispersion), 0.001, 0.005, and 0.01.

Unlike the finite difference and finite element methods, the CVBEM requires no domain discretization and analytical solutions are used in the streamline simulation;

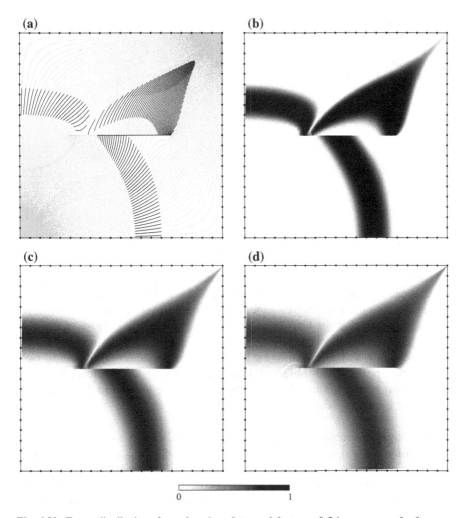

Fig. 6.30 Tracer distributions for a slug size of $\Delta t = 0.2$ at $t = 0.5$ in a quarter of a five-spot pattern containing a fracture of length $L_f = 0.5$ and conductivity $k_f w_f = 10k$. **a** $\alpha_L = 0$ (no dispersion). **b** $\alpha_L = 0.001$. **c** $\alpha_L = 0.005$. **d** $\alpha_L = 0.01$

thus the results are devoid of numerical dispersion. The dispersive behavior shown in Fig. 6.30 is exclusively because of physical (hydrodynamic) dispersion.

Without hydrodynamic dispersion, the leading and trailing interfaces between the tracer and tracer-free fluids are sharp (Fig. 6.30a). With dispersion, the leading and trailing interfaces become indistinct, as seen in Fig. 6.30b–d. As the value of α_L increases, the tracer spreads deeper towards the native fluid, which causes earlier breakthrough of the tracer. At $t = 0.5$, the leading interface without dispersion does

Fig. 6.31 Effluent
tracer-slug concentration for
a slug size of $\Delta t = 0.2$ with
different values of α_L in a
quarter of a five-spot pattern
containing a fracture

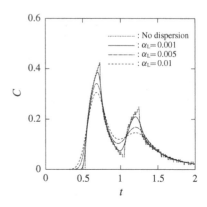

not reach the sink (Fig. 6.30a), whereas it does for the dispersive cases as shown in
Fig. 6.30b–d.

With the same argument as in the previous tasks, the effluent tracer-slug concentration $C(t)$ can be expressed by Eq. 6.60 regardless of dispersivity. The corresponding effluent concentration curves for various dispersivities are shown in Fig. 6.31.

It is confirmed that the breakthrough times become earlier because of the dispersive process, and indeed smaller than $t = 0.5$, as is deduced from the tracer distributions at $t = 0.5$ shown in Fig. 6.30b–d. The characteristic feature of two concentration peaks is still observed. However, for larger values of α_L, the peak concentration becomes lower. Such information may be utilized in the tracer test analysis for evaluating dispersivity.

For Motivating Problem 8, the applicability of the CVBEM is extended to tracer flow with hydrodynamic dispersion in the presence of a fracture. The streamline simulation enables us to use analytical solutions for one-dimensional flow along the streamlines tracked by the CVBEM. The developed scheme is general and applicable to convection-dispersion processes with randomly distributed multiple fractures.

Example 6.3 Let us consider a tracer slug test in a quarter of a five-spot pattern containing multiple fractures. The fracture image is the same as that in Example 6.2; 100 randomly distributed fractures parallel to the x axis with lengths drawn from the normal distribution $N(0.08, 0.03^2)$. The fracture conductivity is $k_f w_f = 10k$.

Except for the fracture configurations, the problem settings and the CVBEM settings are the same as those for Motivating Problem 8. From 201 equiangular points around the wellbore, streamlines are tracked by the stream function method. The longitudinal dispersivity α_L of 0.005 is assumed. Figure 6.32 shows the tracer distributions at different times: $t = 0.1, 0.3, 0.5,$ and 0.7.

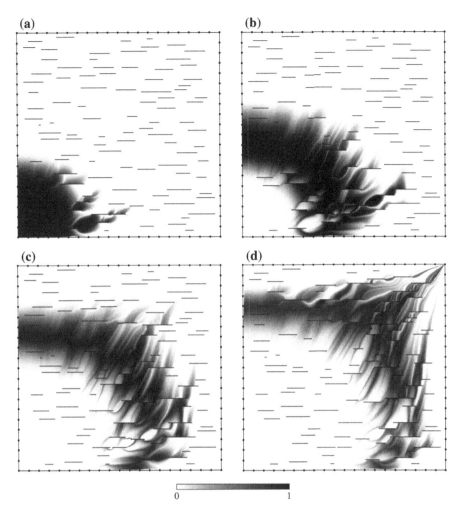

Fig. 6.32 Tracer distributions with dispersivity $\alpha_L = 0.005$ for a slug size of $\Delta t = 0.2$ in a quarter of a five-spot pattern containing 100 randomly distributed fractures with lengths drawn from the normal distribution $N(0.08, 0.03^2)$. **a** $t = 0.1$. **b** $t = 0.3$. **c** $t = 0.5$. **d** $t = 0.7$

Tracer flow through multiple fractures is appropriately simulated. As time proceeds and the tracer advances towards the sink, the tracer distribution becomes more dissipated. Such detailed observations can be made because the CVBEM with the streamline simulation is free from numerical dispersion. The sound mathematical foundations, provided by complex analysis, ensure the computational stability and accuracy of the numerical scheme.

Appendix A
Vector Operator

A.1 Cartesian Coordinates

The differential operator ∇ is given by[1]

$$\nabla = \frac{\partial}{\partial x}\mathbf{i_x} + \frac{\partial}{\partial y}\mathbf{i_y} + \frac{\partial}{\partial z}\mathbf{i_z} \tag{A.1}$$

where $\mathbf{i_x}$, $\mathbf{i_y}$, and $\mathbf{i_z}$ are the unit vectors in Cartesian coordinates.

Gradient (∇)

For a scalar function f, its gradient is defined by

$$\nabla f = \frac{\partial f}{\partial x}\mathbf{i_x} + \frac{\partial f}{\partial y}\mathbf{i_y} + \frac{\partial f}{\partial z}\mathbf{i_z} \tag{A.2}$$

where f is differentiable.

Divergence ($\nabla \cdot$)

For a vector function $\mathbf{v} = (v_x, v_y, v_z)$, its divergence is defined by

$$\nabla \cdot \mathbf{v} = \left(\frac{\partial}{\partial x}\mathbf{i_x} + \frac{\partial}{\partial y}\mathbf{i_y} + \frac{\partial}{\partial z}\mathbf{i_z}\right) \cdot \left(v_x\mathbf{i_x} + v_y\mathbf{i_y} + v_z\mathbf{i_z}\right)$$

$$= \frac{\partial v_x}{\partial x} + \frac{\partial v_y}{\partial y} + \frac{\partial v_z}{\partial z} \tag{A.3}$$

where $\mathbf{v}$ is differentiable.

---

[1] In this appendix, the symbol z is used for a vertical coordinate.

© Springer International Publishing Switzerland 2015
K. Sato, *Complex Analysis for Practical Engineering*,
DOI 10.1007/978-3-319-13063-7

Curl ($\nabla \times$)

For a vector function $\mathbf{v} = (v_x, v_y, v_z)$, its curl (or rotation) is defined by

$$\nabla \times \mathbf{v} = \begin{vmatrix} \mathbf{i_x} & \mathbf{i_y} & \mathbf{i_z} \\ \dfrac{\partial}{\partial x} & \dfrac{\partial}{\partial y} & \dfrac{\partial}{\partial z} \\ v_x & v_y & v_z \end{vmatrix}$$

$$= \left(\frac{\partial v_z}{\partial y} - \frac{\partial v_y}{\partial z} \right) \mathbf{i_x} + \left(\frac{\partial v_x}{\partial z} - \frac{\partial v_z}{\partial x} \right) \mathbf{i_y} + \left(\frac{\partial v_y}{\partial x} - \frac{\partial v_x}{\partial y} \right) \mathbf{i_z} \qquad (A.4)$$

where $\mathbf{v}$ is differentiable. In two dimensions, since $\mathbf{v} = (v_x, v_y, 0)$, $v_x = v_x(x, y)$, and $v_y = v_y(x, y)$, its curl becomes

$$\nabla \times \mathbf{v} = \begin{vmatrix} \mathbf{i_x} & \mathbf{i_y} & \mathbf{i_z} \\ \dfrac{\partial}{\partial x} & \dfrac{\partial}{\partial y} & \dfrac{\partial}{\partial z} \\ v_x & v_y & 0 \end{vmatrix} = \left(\frac{\partial v_y}{\partial x} - \frac{\partial v_x}{\partial y} \right) \mathbf{i_z} \qquad (A.5)$$

Laplacian (∇^2)

For a scalar function f, its Laplacian is defined by

$$\nabla^2 f = \frac{\partial^2 f}{\partial x^2} + \frac{\partial^2 f}{\partial y^2} + \frac{\partial^2 f}{\partial z^2} \qquad (A.6)$$

where f is twice differentiable.

A.2 Cylindrical Coordinates

Cartesian xyz coordinates are related to cylindrical $r\theta z$ coordinates as

$$\begin{cases} x = r \cos \theta \\ y = r \sin \theta \\ z = z \end{cases} \qquad (A.7)$$

with which Eq. A.1 is converted to

$$\nabla = \frac{\partial}{\partial r} \mathbf{i_r} + \frac{1}{r} \frac{\partial}{\partial \theta} \mathbf{i_\theta} + \frac{\partial}{\partial z} \mathbf{i_z} \qquad (A.8)$$

where $\mathbf{i_r}$, $\mathbf{i_\theta}$, and $\mathbf{i_z}$ are the unit vectors in cylindrical coordinates.

Gradient (∇)

For a scalar function f, its gradient is defined by

$$\nabla f = \frac{\partial f}{\partial r}\mathbf{i_r} + \frac{1}{r}\frac{\partial f}{\partial \theta}\mathbf{i}_\theta + \frac{\partial f}{\partial z}\mathbf{i_z} \tag{A.9}$$

where f is differentiable.

Divergence ($\nabla \cdot$)

For a vector function $\mathbf{v} = (v_r, v_\theta, v_z)$, its divergence is defined by

$$\nabla \cdot \mathbf{v} = \frac{1}{r}\frac{\partial (rv_r)}{\partial r} + \frac{1}{r}\frac{\partial v_\theta}{\partial \theta} + \frac{\partial v_z}{\partial z} \tag{A.10}$$

where $\mathbf{v}$ is differentiable.

Curl ($\nabla \times$)

For a vector function $\mathbf{v} = (v_r, v_\theta, v_z)$, its curl (or rotation) is defined by

$$\nabla \times \mathbf{v} = \left(\frac{1}{r}\frac{\partial v_z}{\partial \theta} - \frac{\partial v_\theta}{\partial z}\right)\mathbf{i_r} + \left(\frac{\partial v_r}{\partial z} - \frac{\partial v_z}{\partial r}\right)\mathbf{i}_\theta + \left(\frac{1}{r}\frac{\partial (rv_\theta)}{\partial r} - \frac{1}{r}\frac{\partial v_r}{\partial \theta}\right)\mathbf{i_z} \tag{A.11}$$

where $\mathbf{v}$ is differentiable.

Laplacian (∇^2)

For a scalar function f, its Laplacian is defined by

$$\nabla^2 f = \frac{1}{r}\frac{\partial}{\partial r}\left(r\frac{\partial f}{\partial r}\right) + \frac{1}{r^2}\frac{\partial^2 f}{\partial \theta^2} + \frac{\partial^2 f}{\partial z^2} \tag{A.12}$$

where f is twice differentiable.

Appendix B
Relevant Theorems

B.1 Green's Theorem

Theorem B.1 (Green's theorem) *If $f(x, y)$ and $g(x, y)$ are continuous and have continuous partial derivatives within a region R and on a closed contour C, then*

$$\iint_R \left(\frac{\partial f}{\partial x} - \frac{\partial g}{\partial y} \right) dxdy = \oint_C (gdx + fdy) \tag{B.1}$$

where C is in the counterclockwise direction.

Proof Suppose C is a simple closed curve and any straight line parallel to the x and y axes cuts C in at most two points, as shown in Fig. B.1a. Let ALB and BUA be curves respectively defined by $y = y_1(x)$ and $y = y_2(x)$, where $a \le x \le b$, from $A = (a, y_1(a)) = (a, y_2(a))$ to $B = (b, y_1(b)) = (b, y_2(b))$, then

$$\oint_C gdx = \int_{ALB} gdx + \int_{BUA} gdx$$

$$= \int_a^b g(x, y_1(x))dx + \int_b^a g(x, y_2(x))dx$$

$$= \int_a^b [g(x, y_1(x)) - g(x, y_2(x))]\, dx$$

© Springer International Publishing Switzerland 2015
K. Sato, *Complex Analysis for Practical Engineering*,
DOI 10.1007/978-3-319-13063-7

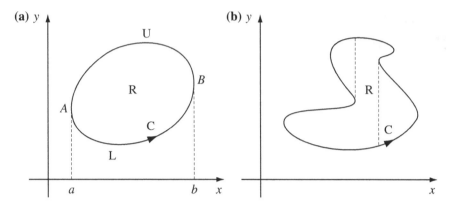

Fig. B.1 Paths of integration. **a** Simple shape. **b** General shape

$$= \int_a^b \left[\int_{y_2(x)}^{y_1(x)} \frac{\partial g(x, y)}{\partial y} dy \right] dx = - \int_a^b \int_{y_1(x)}^{y_2(x)} \frac{\partial g(x, y)}{\partial y} dy dx$$

$$= - \iint_R \frac{\partial g}{\partial y} dx dy \tag{B.2}$$

In a similar way, it follows that

$$\oint_C f dy = \iint_R \frac{\partial f}{\partial x} dx dy \tag{B.3}$$

Adding these equations yields Eq. B.1.

When C is a closed curve in which straight lines parallel to the x and y axes cut C in more than two points, as shown in Fig. B.1b, the region R is divided into subregions, each of which has the property that any straight line parallel to the x and y axes cuts the boundary in at most two points. Then, applying Eq. B.1 to each subregion and adding the results gives Green's theorem. Note that the integrals over the lines introduced for subdivision cancel because the integrations are in both directions. □

B.2 Cauchy's Integral Theorem

Theorem B.2 (Cauchy's integral theorem) *If $f(z)$ is analytic at all points interior to and on a simple closed contour C, then*

$$\oint_C f(z) dz = 0 \tag{B.4}$$

Fig. B.2 Paths of
integration along triangles

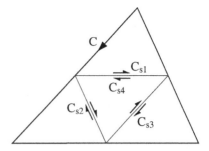

Proof Let us consider a triangle bounded by C, as shown in Fig. B.2. Joining the
midpoints of the sides and subdividing the triangle forms four congruent triangles,
bounded by positively oriented contours C_{s1}, C_{s2}, C_{s3}, and C_{s4}, respectively.

The integrals over the lines introduced for subdivision cancel because the integrations are in both directions, and it follows that

$$\oint_C f(z)dz = \sum_{j=1}^{4} \oint_{C_{sj}} f(z)dz \tag{B.5}$$

Let C_1 be selected from C_{sj} so that the corresponding integral is the largest in absolute
value. Then, by the generalized triangle inequality

$$\left| \oint_C f(z)dz \right| \le \sum_{j=1}^{4} \left| \oint_{C_{sj}} f(z)dz \right| \le 4 \left| \oint_{C_1} f(z)dz \right| \tag{B.6}$$

Subdivide the triangle bounded by C_1 as before and select a contour C_2 so that
the corresponding integral is the largest in absolute value. Then, it follows that

$$\left| \oint_C f(z)dz \right| \le 4 \left| \oint_{C_1} f(z)dz \right| \le 4^2 \left| \oint_{C_2} f(z)dz \right| \tag{B.7}$$

Continuing this procedure yields

$$\left| \oint_C f(z)dz \right| \le 4^n \left| \oint_{C_n} f(z)dz \right| \tag{B.8}$$

Let z_0 be the point that lies in all these triangles. Since $f(z)$ is analytic at z_0, Eq. 2.74 holds

$$f(z) = f(z_0) + f'(z_0)(z - z_0) + \varepsilon(z - z_0) \tag{B.9}$$

where for any positive number ε_0, some positive number δ can be found such that $|\varepsilon| < \varepsilon_0$ whenever $|z - z_0| < \delta$. Integrating both sides over C_n gives

$$\oint_{C_n} f(z)\mathrm{d}z = \oint_{C_n} f(z_0)\mathrm{d}z + \oint_{C_n} f'(z_0)(z - z_0)\mathrm{d}z + \oint_{C_n} \varepsilon(z - z_0)\mathrm{d}z \tag{B.10}$$

The first two integrals on the right-hand side vanish, since the integrands have continuous derivatives, to which Cauchy's integral theorem proved in Sect. 4.2.4 can be applied. Hence, it follows that

$$\oint_{C_n} f(z)\mathrm{d}z = \oint_{C_n} \varepsilon(z - z_0)\mathrm{d}z \tag{B.11}$$

Let L be the length of C, then the length of C_n is $L_n = L/2^n$. If z is on the triangle bounded by C_n, the distance between z and z_0 is less than the perimeter of the triangle, that is, $|z - z_0| < L/2^n$. From Eq. B.11 and the ML inequality, it follows that

$$\left| \oint_{C_n} f(z)\mathrm{d}z \right| = \left| \oint_{C_n} \varepsilon(z - z_0)\mathrm{d}z \right| \leq \varepsilon_0 \frac{L}{2^n} \frac{L}{2^n} = \varepsilon_0 \frac{L^2}{4^n} \tag{B.12}$$

Substituting Eq. B.12 into Eq. B.8 yields

$$\left| \oint_{C} f(z)\mathrm{d}z \right| \leq \varepsilon_0 L^2 \tag{B.13}$$

Since ε_0 can be arbitrarily small, the value of the integral must be zero and the proof is complete for triangular contours.

When C is a polygonal contour, the polygon is subdivided into a finite number of triangles. The integral over each triangle is zero, and the sum of these integrals is equal to the integral over the polygonal contour C, which completes the proof for polygonal contours. For a general simple closed contour, the contour is approximated by a sufficiently close polygonal contour. $\qquad\square$

Appendix C
Conservative Field

C.1 Existence of Potential Functions

Let us consider the case when the curl of a two-dimensional vector function $\mathbf{v} = v_x\mathbf{i_x} + v_y\mathbf{i_y}$ vanishes, that is,

$$\nabla \times \mathbf{v} = \mathbf{0} \tag{C.1}$$

which is equivalent to

$$\frac{\partial v_y}{\partial x} = \frac{\partial v_x}{\partial y} \tag{C.2}$$

where Eq. A.5 is used.

Now, define a scalar function f for a fixed point (x_0, y_0) as

$$f = \int_{x_0}^{x} v_x(x, y)\mathrm{d}x + \int_{y_0}^{y} v_y(x_0, y)\mathrm{d}y \tag{C.3}$$

Then, the gradient of f becomes

$$\nabla f = v_x(x, y)\mathbf{i_x} + \left[\int_{x_0}^{x} \frac{\partial}{\partial y}v_x(x, y)\mathrm{d}x + v_y(x_0, y) \right]\mathbf{i_y} \tag{C.4}$$

From Eq. C.2, it follows that

$$\int_{x_0}^{x} \frac{\partial}{\partial y}v_x(x, y)\mathrm{d}x + v_y(x_0, y) = \int_{x_0}^{x} \frac{\partial}{\partial x}v_y(x, y)\mathrm{d}x + v_y(x_0, y)$$

$$= v_y(x, y)\Big|_{x_0}^{x} + v_y(x_0, y)$$

$$= v_y(x, y) \tag{C.5}$$

© Springer International Publishing Switzerland 2015
K. Sato, *Complex Analysis for Practical Engineering*,
DOI 10.1007/978-3-319-13063-7

Substituting Eq. C.5 into Eq. C.4 yields

$$\nabla f = v_x \mathbf{i_x} + v_y \mathbf{i_y} = \mathbf{v} \tag{C.6}$$

Therefore, when $\nabla \times \mathbf{v} = \mathbf{0}$, there must exist a scalar function f, such that

$$\mathbf{v} = \nabla f = \frac{\partial f}{\partial x}\mathbf{i_x} + \frac{\partial f}{\partial y}\mathbf{i_y} \tag{C.7}$$

The function f is called the potential function of $\mathbf{v}$, and the vector $\mathbf{v}$ and its vector field are said to be conservative.

C.2 Path Independence and Exactness

Let C be a curve defined by $\mathbf{r}(t) = x(t)\mathbf{i_x} + y(t)\mathbf{i_y}$, where $a \le t \le b$, from $A = (x(a), y(a))$ to $B = (x(b), y(b))$. Then, the line integral of the conservative vector $\mathbf{v}$ along C becomes

$$\int_C \mathbf{v} \cdot d\mathbf{r} = \int_C \left(\frac{\partial f}{\partial x}dx + \frac{\partial f}{\partial y}dy \right)$$

$$= \int_a^b \left(\frac{\partial f}{\partial x}\frac{dx}{dt} + \frac{\partial f}{\partial y}\frac{dy}{dt} \right) dt$$

$$= \int_a^b \frac{df}{dt}dt = f(x(b), y(b)) - f(x(a), y(a))$$

$$= f(B) - f(A) \tag{C.8}$$

where the differential form of f is given by

$$df = \frac{\partial f}{\partial x}dx + \frac{\partial f}{\partial y}dy \tag{C.9}$$

which is called exact.

Equation C.8 implies that if $\mathbf{v}$ is conservative the line integral of $\mathbf{v}$ is path independent. In particular, when C is a closed curve; that is, $A = B$, it can be shown that

$$\oint_C \mathbf{v} \cdot d\mathbf{r} = 0 \tag{C.10}$$

Equations C.1, C.7, C.9, and C.10 are mutually equivalent.

Appendix D
Coefficients in Singularity Programming

D.1 Coefficients of Fracture Parameters to Boundary Nodes

When the velocity potential is prescribed at the boundary node ζ_k, the imaginary part of $2\pi i \Omega_{fl}(\zeta_k)$ is used with the boundary conditions, Eq. 5.158. Thus, the coefficients of the lth fracture to the boundary node are given by

$$B_{1nl} = 2\pi \, \mathrm{Re} \left[\frac{1}{\chi_l^n} \right] \tag{D.1}$$

$$B_{2jl} = 2\pi \, \mathrm{Re} \left[\frac{\chi_{pj}}{\chi_l - \chi_{pj}} + \frac{\overline{\chi_{pj}}}{\chi_l - \overline{\chi_{pj}}} \right] \tag{D.2}$$

$$B_{3jl} = 2\pi \, \mathrm{Im} \left[-\frac{\chi_{pj}}{\chi_l - \chi_{pj}} + \frac{\overline{\chi_{pj}}}{\chi_l - \overline{\chi_{pj}}} \right] \tag{D.3}$$

which correspond to a_{-nl}, b_{jl}, and c_{jl}, respectively.

When the stream function is prescribed at the boundary node ζ_k, the real part of $2\pi i \Omega_{fl}(\zeta_k)$ is used with the boundary conditions, Eq. 5.159. Thus, the coefficients of the lth fracture to the boundary node are given by

$$B_{1nl} = -2\pi \, \mathrm{Im} \left[\frac{1}{\chi_l^n} \right] \tag{D.4}$$

$$B_{2jl} = -2\pi \, \mathrm{Im} \left[\frac{\chi_{pj}}{\chi_l - \chi_{pj}} + \frac{\overline{\chi_{pj}}}{\chi_l - \overline{\chi_{pj}}} \right] \tag{D.5}$$

$$B_{3jl} = -2\pi \, \mathrm{Re} \left[\frac{\chi_{pj}}{\chi_l - \chi_{pj}} - \frac{\overline{\chi_{pj}}}{\chi_l - \overline{\chi_{pj}}} \right] \tag{D.6}$$

which correspond to a_{-nl}, b_{jl}, and c_{jl}, respectively. In all these equations, χ_l is the image of $z = \zeta_k$, given by Eq. 5.151 with z_{1l} and z_{2l} of the lth fracture.

© Springer International Publishing Switzerland 2015
K. Sato, *Complex Analysis for Practical Engineering*,
DOI 10.1007/978-3-319-13063-7

D.2 Coefficients to Fractures

In the presence of multiple fractures, as a modification of Eq. 6.17, the boundary condition along the mth fracture becomes

$$
\Psi_{\mathrm{f}m}(z^+) - \Psi_{\mathrm{f}m}(z^-) - \frac{k_{\mathrm{f}m}w_{\mathrm{f}m}}{k}\left(\frac{d\tilde{\Phi}^{\mathrm{ns}}}{ds_m}(z) + \sum_{l=1}^{n_{\mathrm{f}}} \frac{d\Phi_{\mathrm{f}l}}{ds_m}(z) \right)
$$

$$
= \frac{k_{\mathrm{f}m}w_{\mathrm{f}m}}{k} \sum_{j=1}^{n_{\mathrm{w}}} \frac{d\Phi_{\mathrm{w}j}}{ds_m}(z) \tag{D.7}
$$

where $k_{\mathrm{f}m}w_{\mathrm{f}m}$ is the conductivity of, s_m is the coordinate along, and z is the selected point on the mth fracture.

From the fracture singular solution, Eq. 5.155, the difference between the values of Ψ on the $+$ and $-$ sides of the mth fracture at z becomes

$$
\Psi_{\mathrm{f}m}(z^+) - \Psi_{\mathrm{f}m}(z^-) = \sum_{n=1}^{n_{\mathrm{L}}} \mathrm{Im}\left[\frac{1}{(\chi_m^+)^n} - \frac{1}{(\chi_m^-)^n} \right] a_{-nm}
$$

$$
+ \sum_{j=1}^{n_{\mathrm{p}}} \mathrm{Im}\left[\frac{\chi_{\mathrm{p}j}}{\chi_m^+ - \chi_{\mathrm{p}j}} + \frac{\overline{\chi_{\mathrm{p}j}}}{\chi_m^+ - \overline{\chi_{\mathrm{p}j}}} - \frac{\chi_{\mathrm{p}j}}{\chi_m^- - \chi_{\mathrm{p}j}} - \frac{\overline{\chi_{\mathrm{p}j}}}{\chi_m^- - \overline{\chi_{\mathrm{p}j}}} \right] b_{jm}
$$

$$
+ \sum_{j=2}^{n_{\mathrm{p}}-1} \mathrm{Re}\left[\frac{\chi_{\mathrm{p}j}}{\chi_m^+ - \chi_{\mathrm{p}j}} - \frac{\overline{\chi_{\mathrm{p}j}}}{\chi_m^+ - \overline{\chi_{\mathrm{p}j}}} - \frac{\chi_{\mathrm{p}j}}{\chi_m^- - \chi_{\mathrm{p}j}} + \frac{\overline{\chi_{\mathrm{p}j}}}{\chi_m^- - \overline{\chi_{\mathrm{p}j}}} \right] c_{jm} \tag{D.8}
$$

where $\chi_m^{\pm}$ is given by Eq. 5.151 for z with z_{1m} and z_{2m} of the mth fracture.

The derivatives of $\tilde{\Phi}^{\mathrm{ns}}(z)$ and $\Phi_{\mathrm{f}l}(z)$ with respect to s_m can be obtained as

$$
\frac{d\tilde{\Phi}^{\mathrm{ns}}}{ds_m}(z) + \sum_{l=1}^{n_{\mathrm{f}}} \frac{d\Phi_{\mathrm{f}l}}{ds_m}(z) = \mathrm{Re}\left[\frac{d\tilde{\Omega}^{\mathrm{ns}}}{d\varsigma_m}(z) + \sum_{l=1}^{n_{\mathrm{f}}} \frac{d\Omega_{\mathrm{f}l}}{d\varsigma_m}(z) \right] \tag{D.9}
$$

where ς_m is the complex variable with s_m as its real part, and is given by

$$
\varsigma_m = \frac{|z_{2m} - z_{1m}|}{z_{2m} - z_{1m}} z = \frac{L_{\mathrm{f}m}}{z_{2m} - z_{1m}} z \tag{D.10}
$$

where $L_{\mathrm{f}m}$ is the length of the mth fracture.

The influence of the non-singular solution to the gradient of the velocity potential along the mth fracture is given by

$$
\frac{d\tilde{\Phi}^{\mathrm{ns}}}{ds_m}(z) = \mathrm{Re}\left[\frac{d\tilde{\Omega}^{\mathrm{ns}}}{d\varsigma_m}(z) \right] = \mathrm{Re}\left[\frac{dz}{d\varsigma_m} \frac{d\tilde{\Omega}^{\mathrm{ns}}}{dz}(z) \right]
$$

$$= \mathrm{Re}\left[\frac{z_{2m} - z_{1m}}{L_{fm}} \frac{d\tilde{\Omega}^{ns}}{dz}(z)\right] \tag{D.11}$$

where $dz/d\varsigma_m$ is obtained from Eq. D.10. Substituting Eq. 6.38 into Eq. D.11 results in

$$\frac{d\tilde{\Phi}^{ns}}{ds_m}(z) = \sum_{j=1}^{n_b} \mathrm{Re}\left[\frac{z_{2m} - z_{1m}}{2\pi i L_{fm}} \frac{\Omega_{j+1}^{ns} - \Omega_j^{ns}}{\varsigma_{j+1} - \varsigma_j} h_j\right]$$

$$= \sum_{j=1}^{n_b} \mathrm{Re}\left[\frac{z_{2m} - z_{1m}}{2\pi i L_{fm}} \frac{\Phi_{j+1}^{ns} - \Phi_j^{ns}}{\varsigma_{j+1} - \varsigma_j} h_j\right]$$

$$- \sum_{j=1}^{n_b} \mathrm{Im}\left[\frac{z_{2m} - z_{1m}}{2\pi i L_{fm}} \frac{\Psi_{j+1}^{ns} - \Psi_j^{ns}}{\varsigma_{j+1} - \varsigma_j} h_j\right] \tag{D.12}$$

where h_j is given by

$$h_j = \ln\frac{\varsigma_{j+1} - z}{\varsigma_j - z} \tag{D.13}$$

In a similar way, the influence of the lth fracture to the gradient of the velocity potential along the mth fracture is given by

$$\frac{d\Phi_{fl}}{ds_m}(z) = \mathrm{Re}\left[\frac{d\Omega_{fl}}{d\varsigma_m}(z)\right] = \mathrm{Re}\left[\frac{dz}{d\varsigma_m} \frac{d\chi_l}{dz} \frac{d\Omega_{fl}}{d\chi_l}(\chi)\right]$$

$$= \mathrm{Re}\left[\frac{z_{2m} - z_{1m}}{L_{fm}} \frac{d\chi_l}{dz} \frac{d\Omega_{fl}}{d\chi_l}(\chi)\right] \tag{D.14}$$

where $d\chi_l/dz$ and $d\Omega_{fl}(\chi)/d\chi_l$ are respectively given by Eqs. 6.41 and 6.42, and it follows that

$$\frac{d\Phi_{fl}}{ds_m} = -\sum_{n=1}^{n_L} \mathrm{Re}\left[\frac{4\chi_l^2(z_{2m} - z_{1m})}{L_{fm}(\chi_l^2 - 1)(z_{2l} - z_{1l})} \frac{n}{\chi_l^{n+1}}\right] a_{-nl}$$

$$- \sum_{j=1}^{n_p} \mathrm{Re}\left[\frac{4\chi_l^2(z_{2m} - z_{1m})}{L_{fm}(\chi_l^2 - 1)(z_{2l} - z_{1l})}\left\{\frac{\chi_{pj}}{(\chi_l - \chi_{pj})^2} + \frac{\overline{\chi_{pj}}}{(\chi_l - \overline{\chi_{pj}})^2}\right\}\right] b_{jl}$$

$$+ \sum_{j=2}^{n_p-1} \mathrm{Im}\left[\frac{4\chi_l^2(z_{2m} - z_{1m})}{L_{fm}(\chi_l^2 - 1)(z_{2l} - z_{1l})}\left\{\frac{\chi_{pj}}{(\chi_l - \chi_{pj})^2} - \frac{\overline{\chi_{pj}}}{(\chi_l - \overline{\chi_{pj}})^2}\right\}\right] c_{jl}$$

$$\tag{D.15}$$

On the right-hand side of Eq. D.7, the influence of the n_w-source/sink system on the gradient of the velocity potential along the mth fracture is explicit, the value of which is given by

$$\sum_{j=1}^{n_w} \frac{d\Phi_{wj}}{ds_m}(z) = \sum_{j=1}^{n_w} \mathrm{Re}\left[\frac{d\Omega_{wj}}{ds_m}(z)\right] = \sum_{j=1}^{n_w} \mathrm{Re}\left[\frac{dz}{ds_m}\frac{d\Omega_{wj}}{dz}(z)\right]$$

$$= \sum_{j=1}^{n_w} \mathrm{Re}\left[\frac{z_{2m} - z_{1m}}{L_{fm}}\frac{d\Omega_{wj}}{dz}(z)\right] \tag{D.16}$$

where $d\Omega_{wj}(z)/dz$ is given by Eq. 6.39, and it follows that

$$\sum_{j=1}^{n_w} \frac{d\Phi_{wj}}{ds_m}(z) = \sum_{j=1}^{n_w} \mathrm{Re}\left[-\frac{q_{wj}}{2\pi h L_{fm}}\frac{z_{2m} - z_{1m}}{z - z_{wj}}\right] \tag{D.17}$$

Substituting Eqs. D.8, D.12, D.15, and D.17 into the boundary condition along the mth fracture, Eq. D.7, yields the desired simultaneous linear equations, and the coefficients in Eqs. 5.162 and 6.22 can be obtained as follows.

Coefficients of the jth Boundary Node to the mth Fracture

Substituting Eq. D.12 into Eq. D.7 yields the coefficients of the jth boundary node to the mth fracture as

$$C_{1j} = -\mathrm{Re}\left[\frac{z_{2m} - z_{1m}}{2\pi i L_{fm}}\left(\frac{h_{j-1}}{\zeta_j - \zeta_{j-1}} - \frac{h_j}{\zeta_{j+1} - \zeta_j}\right)\right]\frac{k_{fm}w_{fm}}{k} \tag{D.18}$$

$$C_{2j} = \mathrm{Im}\left[\frac{z_{2m} - z_{1m}}{2\pi i L_{fm}}\left(\frac{h_{j-1}}{\zeta_j - \zeta_{j-1}} - \frac{h_j}{\zeta_{j+1} - \zeta_j}\right)\right]\frac{k_{fm}w_{fm}}{k} \tag{D.19}$$

which correspond to Φ_j^{ns} and Ψ_j^{ns}, respectively.

Coefficients of the mth Fracture to Itself

Substituting Eqs. D.8 and D.15 into Eq. D.7 yields the coefficients of the mth fracture to itself as

$$D_{1n} = \mathrm{Im}\left[\frac{1}{(\chi_m^+)^n} - \frac{1}{(\chi_m^-)^n}\right] + \mathrm{Re}\left[\frac{4n}{L_{fm}\chi_m^{n-1}(\chi_m^2 - 1)}\right]\frac{k_{fm}w_{fm}}{k} \tag{D.20}$$

$$D_{2j} = \mathrm{Im}\left[\frac{\chi_{pj}}{\chi_m^+ - \chi_{pj}} + \frac{\overline{\chi_{pj}}}{\chi_m^+ - \overline{\chi_{pj}}} - \frac{\chi_{pj}}{\chi_m^- - \chi_{pj}} - \frac{\overline{\chi_{pj}}}{\chi_m^- - \overline{\chi_{pj}}}\right]$$

$$+ \mathrm{Re}\left[\frac{4\chi_m^2}{L_{fm}(\chi_m^2 - 1)}\left\{\frac{\chi_{pj}}{(\chi_m - \chi_{pj})^2} + \frac{\overline{\chi_{pj}}}{(\chi_m - \overline{\chi_{pj}})^2}\right\}\right]\frac{k_{fm}w_{fm}}{k} \quad (D.21)$$

$$D_{3j} = \mathrm{Re}\left[\frac{\chi_{pj}}{\chi_m^+ - \chi_{pj}} - \frac{\overline{\chi_{pj}}}{\chi_m^+ - \overline{\chi_{pj}}} - \frac{\chi_{pj}}{\chi_m^- - \chi_{pj}} + \frac{\overline{\chi_{pj}}}{\chi_m^- - \overline{\chi_{pj}}}\right]$$

$$- \mathrm{Im}\left[\frac{4\chi_m^2}{L_{fm}(\chi_m^2 - 1)}\left\{\frac{\chi_{pj}}{(\chi_m - \chi_{pj})^2} - \frac{\overline{\chi_{pj}}}{(\chi_m - \overline{\chi_{pj}})^2}\right\}\right]\frac{k_{fm}w_{fm}}{k} \quad (D.22)$$

which correspond to a_{-nm}, b_{jm}, and c_{jm}, respectively.

Coefficients of the *l*th Fracture to the *m*th Fracture

Substituting Eq. D.15 into Eq. D.7 yields the coefficients of the *l*th fracture to the *m*th fracture as

$$E_{1nl} = \mathrm{Re}\left[\frac{4n(z_{2m} - z_{1m})}{L_{fm}\chi_l^{n-1}(\chi_l^2 - 1)(z_{2l} - z_{1l})}\right]\frac{k_{fm}w_{fm}}{k} \quad (D.23)$$

$$E_{2jl} = \mathrm{Re}\left[\frac{4\chi_l^2(z_{2m} - z_{1m})}{L_{fm}(\chi_l^2 - 1)(z_{2l} - z_{1l})}\left\{\frac{\chi_{pj}}{(\chi_l - \chi_{pj})^2} + \frac{\overline{\chi_{pj}}}{(\chi_l - \overline{\chi_{pj}})^2}\right\}\right]\frac{k_{fm}w_{fm}}{k}$$

$$\quad (D.24)$$

$$E_{3jl} = -\mathrm{Im}\left[\frac{4\chi_l^2(z_{2m} - z_{1m})}{L_{fm}(\chi_l^2 - 1)(z_{2l} - z_{1l})}\left\{\frac{\chi_{pj}}{(\chi_l - \chi_{pj})^2} - \frac{\overline{\chi_{pj}}}{(\chi_l - \overline{\chi_{pj}})^2}\right\}\right]\frac{k_{fm}w_{fm}}{k}$$

$$\quad (D.25)$$

which correspond to a_{-nl}, b_{jl}, and c_{jl}, respectively.

Constant of the *n*w-source/sink System to the *m*th Fracture

Substituting Eq. D.17 into Eq. D.7 yields the constant of n_w-source/sink system to the *m*th fracture as

$$F = \sum_{j=1}^{n_w} \mathrm{Re}\left[-\frac{q_{wj}}{2\pi h L_{fm}}\frac{z_{2m} - z_{1m}}{z - z_{wj}}\right]\frac{k_{fm}w_{fm}}{k} \quad (D.26)$$

Appendix E
Example Computation of the CVBEM

E.1 Dual-Variable Equivalence

Let us consider uniform inclined flow given by

$$\Omega(z) = -e^{-(\pi/3)i}z \tag{E.1}$$

through a square (1×1) flow domain, as shown in Fig. E.1. The velocity potential and stream function are respectively given by

$$\begin{cases} \Phi = -(1/2)x - (\sqrt{3}/2)y \\ \Psi = (\sqrt{3}/2)x - (1/2)y \end{cases} \tag{E.2}$$

which are used to prescribe the boundary conditions.

In the CVBEM, a single boundary element for each side of the square domain is sufficient, since the flow is uniform. The domain is modeled with four boundary elements ($n_b = 4$), and the Dirichlet boundary conditions are assigned at the nodal points ζ_1, ζ_2, and ζ_3, whereas the Neumann boundary condition is assigned at ζ_4.

Since the velocity potential is prescribed at the nodal point ζ_1, the approximate boundary value $\tilde{\Phi}(\zeta_1)$ is equated with the prescribed value P_1 and the approximate boundary value $\tilde{\Psi}(\zeta_1)$ with the nodal value Ψ_1:

$$\begin{cases} \tilde{\Phi}(\zeta_1) = P_1 \\ \tilde{\Psi}(\zeta_1) = \Psi_1 \end{cases} \tag{E.3}$$

From Eq. 4.62, it follows

$$2\pi P_1 = \ln \left| \frac{\zeta_2 - \zeta_1}{\zeta_4 - \zeta_1} \right| \Psi_1 + (2\pi - \theta_1)\Phi_1$$

$$+ \, \mathrm{Re}[H_2]\Psi_3 + \mathrm{Im}[H_2]\Phi_3 - \mathrm{Re}[I_2]\Psi_2 - \mathrm{Im}[I_2]\Phi_2$$

$$+ \, \mathrm{Re}[H_3]\Psi_4 + \mathrm{Im}[H_3]\Phi_4 - \mathrm{Re}[I_3]\Psi_3 - \mathrm{Im}[I_3]\Phi_3 \tag{E.4}$$

© Springer International Publishing Switzerland 2015
K. Sato, *Complex Analysis for Practical Engineering*,
DOI 10.1007/978-3-319-13063-7

Fig. E.1 Uniform inclined
flow through a square flow
domain

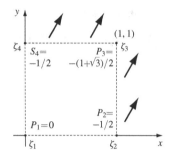

and from Eq. 4.63, it follows

$$0 = \ln \left| \frac{\zeta_2 - \zeta_1}{\zeta_4 - \zeta_1} \right| \Phi_1 + \theta_1 \Psi_1$$

$$+ \operatorname{Re}[H_2]\Phi_3 - \operatorname{Im}[H_2]\Psi_3 - \operatorname{Re}[I_2]\Phi_2 + \operatorname{Im}[I_2]\Psi_2$$

$$+ \operatorname{Re}[H_3]\Phi_4 - \operatorname{Im}[H_3]\Psi_4 - \operatorname{Re}[I_3]\Phi_3 + \operatorname{Im}[I_3]\Psi_3 \qquad (E.5)$$

The coefficients in Eqs. E.4 and E.5, with $\zeta_1 = 0$, $\zeta_2 = 1$, $\zeta_3 = 1 + i$, and $\zeta_4 = i$,
are obtained as

$$\begin{cases} \ln \left| \dfrac{\zeta_2 - \zeta_1}{\zeta_4 - \zeta_1} \right| = \ln |-i| = 0 \\[2mm] \theta_1 = \dfrac{\pi}{2} \\[2mm] H_2 = \dfrac{\zeta_1 - \zeta_2}{\zeta_3 - \zeta_2} \ln \dfrac{\zeta_3 - \zeta_1}{\zeta_2 - \zeta_1} = \dfrac{-1}{i} \ln(1+i) = -\pi/4 + i \ln \sqrt{2} \\[2mm] I_2 = \dfrac{\zeta_1 - \zeta_3}{\zeta_3 - \zeta_2} \ln \dfrac{\zeta_3 - \zeta_1}{\zeta_2 - \zeta_1} \\[2mm] \quad = \dfrac{(-1-i)}{i} \ln(1+i) = -\pi/4 - \ln \sqrt{2} - i(\pi/4 - \ln \sqrt{2}) \\[2mm] H_3 = \dfrac{\zeta_1 - \zeta_3}{\zeta_4 - \zeta_3} \ln \dfrac{\zeta_4 - \zeta_1}{\zeta_3 - \zeta_1} \\[2mm] \quad = \dfrac{(-1-i)}{-1} \ln \dfrac{1+i}{2} = -\pi/4 - \ln \sqrt{2} + i(\pi/4 - \ln \sqrt{2}) \\[2mm] I_3 = \dfrac{\zeta_1 - \zeta_4}{\zeta_4 - \zeta_3} \ln \dfrac{\zeta_4 - \zeta_1}{\zeta_3 - \zeta_1} = \dfrac{-i}{-1} \ln \dfrac{1+i}{2} = -\pi/4 - i \ln \sqrt{2} \end{cases} \qquad (E.6)$$

Substituting these values into Eqs. E.4 and E.5 yields

$$\begin{cases} 2\pi \, P_1 = (3\pi/2)\Phi_1 + \gamma_2\Phi_2 + \gamma_1\Psi_2 + \gamma_3\Phi_3 + \gamma_2\Phi_4 - \gamma_1\Psi_4 \\[1mm] 0 = (\pi/2)\Psi_1 + \gamma_1\Phi_2 - \gamma_2\Psi_2 - \gamma_3\Psi_3 - \gamma_1\Phi_4 - \gamma_2\Psi_4 \end{cases} \qquad (E.7)$$

where $\gamma_1 = \pi/4 + \ln \sqrt{2}$, $\gamma_2 = \pi/4 - \ln \sqrt{2}$, and $\gamma_3 = \ln 2$.

In a similar way, the corresponding equations at the nodal points ζ_2, ζ_3, and ζ_4 are obtained as

$$\begin{cases} 2\pi P_2 = \gamma_2\Phi_1 - \gamma_1\Psi_1 + (3\pi/2)\Phi_2 + \gamma_2\Phi_3 + \gamma_1\Psi_3 + \gamma_3\Phi_4 \\ 0 = -\gamma_1\Phi_1 - \gamma_2\Psi_1 + (\pi/2)\Psi_2 + \gamma_1\Phi_3 - \gamma_2\Psi_3 - \gamma_3\Psi_4 \\ 2\pi P_3 = \gamma_3\Phi_1 + \gamma_2\Phi_2 - \gamma_1\Psi_2 + (3\pi/2)\Phi_3 + \gamma_2\Phi_4 + \gamma_1\Psi_4 \\ 0 = -\gamma_3\Psi_1 - \gamma_1\Phi_2 - \gamma_2\Psi_2 + (\pi/2)\Psi_3 + \gamma_1\Phi_4 - \gamma_2\Psi_4 \\ 0 = \gamma_2\Phi_1 + \gamma_1\Psi_1 + \gamma_3\Phi_2 + \gamma_2\Phi_3 - \gamma_1\Psi_3 - (\pi/2)\Phi_4 \\ -2\pi S_4 = \gamma_1\Phi_1 - \gamma_2\Psi_1 - \gamma_3\Psi_2 - \gamma_1\Phi_3 - \gamma_2\Psi_3 - (3\pi/2)\Psi_4 \end{cases} \qquad \text{(E.8)}$$

Equations E.7 and E.8 form the simultaneous linear equations

$$\begin{pmatrix} 3\pi/2 & 0 & \gamma_2 & \gamma_1 & \gamma_3 & 0 & \gamma_2 & -\gamma_1 \\ 0 & \pi/2 & \gamma_1 & -\gamma_2 & 0 & -\gamma_3 & -\gamma_1 & -\gamma_2 \\ \gamma_2 & -\gamma_1 & 3\pi/2 & 0 & \gamma_2 & \gamma_1 & \gamma_3 & 0 \\ -\gamma_1 & -\gamma_2 & 0 & \pi/2 & \gamma_1 & -\gamma_2 & 0 & -\gamma_3 \\ \gamma_3 & 0 & \gamma_2 & -\gamma_1 & 3\pi/2 & 0 & \gamma_2 & \gamma_1 \\ 0 & -\gamma_3 & -\gamma_1 & -\gamma_2 & 0 & \pi/2 & \gamma_1 & -\gamma_2 \\ \gamma_2 & \gamma_1 & \gamma_3 & 0 & \gamma_2 & -\gamma_1 & -\pi/2 & 0 \\ \gamma_1 & -\gamma_2 & 0 & -\gamma_3 & -\gamma_1 & -\gamma_2 & 0 & -3\pi/2 \end{pmatrix} \begin{pmatrix} \Phi_1 \\ \Psi_1 \\ \Phi_2 \\ \Psi_2 \\ \Phi_3 \\ \Psi_3 \\ \Phi_4 \\ \Psi_4 \end{pmatrix} = \begin{pmatrix} 2\pi P_1 \\ 0 \\ 2\pi P_2 \\ 0 \\ 2\pi P_3 \\ 0 \\ 0 \\ -2\pi S_4 \end{pmatrix}$$

$$\text{(E.9)}$$

the solutions of which are obtained as

$$\begin{pmatrix} \Omega_1 \\ \Omega_2 \\ \Omega_3 \\ \Omega_4 \end{pmatrix} = \begin{pmatrix} \Phi_1 + i\Psi_1 \\ \Phi_2 + i\Psi_2 \\ \Phi_3 + i\Psi_3 \\ \Phi_4 + i\Psi_4 \end{pmatrix} = \begin{pmatrix} 0.0 + 0.0i \\ -0.5000 + 0.8660i \\ -1.366 + 0.3660i \\ -0.8660 - 0.5000i \end{pmatrix} \qquad \text{(E.10)}$$

which are consistent with the analytical solution, Eq. E.2.

E.2 Logarithmic Singularity

The problem considered in Sect. 6.1.4 is solved with a smaller number of $n_b = 8$ to exemplify the treatment of logarithmic singularities more clearly. The boundary DABCD (Fig. 6.13) is represented by arranging eight evenly spaced boundary elements, as shown in Fig. E.2.

Fig. E.2 Branch cuts for the
sink and source at nodal
points

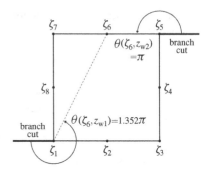

To avoid the logarithmic singularity, as indicated in Fig. 6.14, the nodal point ζ_1 that coincides with z_{w1} is shifted to an expedient node $\zeta_1' = r_{w1} + 0\mathrm{i} = 0.001 + 0\mathrm{i}$ and the nodal point ζ_5 that coincides with z_{w2} is shifted to an expedient node $\zeta_5' = (1 - r_{w2}) + 1\mathrm{i} = 0.999 + 1\mathrm{i}$.

The branch cut from z_{w1} is taken in the negative x direction and that on z_{w2} is taken in the positive x direction, as shown in Fig. E.2, so that the arguments $\theta(\zeta_k, z_{w1})$ and $\theta(\zeta_k, z_{w2})$ can take their principal values without crossing the branch cut. For instance, at the boundary node ζ_6, $\theta(\zeta_6, z_{w1}) = 1.352\pi$ and $\theta(\zeta_6, z_{w2}) = \pi$, and from Eq. 6.5, the boundary value attributed to the singular solution is $\Psi_6^s = -1.352\pi/(2\pi) + \pi/(2\pi) = -0.176$. Then, from Eq. 6.8, the non-singular boundary value is obtained as $S_6 - \Psi_6^s = 0.176$.

In a similar way, the non-singular boundary values at the nodal points ζ_k are obtained as shown in Table E.1.

Table E.1 Non-singular boundary values

Node	S_k	$\theta(\zeta_k, z_{w1})$	$\theta(\zeta_k, z_{w2})$	Ψ_k^s	$S_k - \Psi_k^s$
ζ_1	0.25	π	1.25π	0.125	0.125
ζ_2	0.25	π	1.352π	0.176	0.074
ζ_3	0.25	π	1.5π	0.25	0.0
ζ_4	0.25	1.148π	1.5π	0.176	0.074
ζ_5	0.0	1.25π	π	-0.125	0.125
ζ_6	0.0	1.352π	π	-0.176	0.176
ζ_7	0.0	1.5π	π	-0.25	0.25
ζ_8	0.0	1.5π	1.148π	-0.176	0.176

Fig. E.3 Non-singular (*closed squares*) and prescribed (*open squares*) boundary values with $n_b = 8$ for a quarter of a five-spot pattern

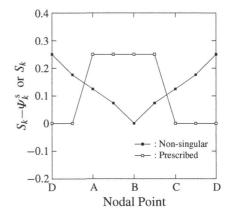

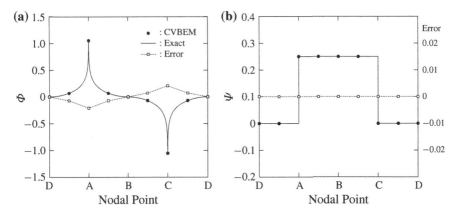

Fig. E.4 Approximate boundary values, $\tilde{\Phi}(\zeta_k)$ and $\tilde{\Psi}(\zeta_k)$ (*black dots*), exact solutions, $\Phi(\zeta_k)$ and $\Psi(\zeta_k)$ (*solid lines*), and computational errors, $\tilde{E}_\Phi(\zeta_k)$ and $\tilde{E}_\Psi(\zeta_k)$ (*open squares*) with $n_b = 8$ for flow in a quarter of a five-spot pattern. **a** Velocity potential. **b** Stream function

Figure E.3 shows the profiles of $S_k - \Psi_k^s$ and S_k. The prescribed (complete) boundary value S_k is not continuous, while the non-singular value $S_k - \Psi_k^s$ is continuous along the Neumann-type boundary DABCD.

Figure E.4 shows the CVBEM result, compared with the analytical solution. The difference in the velocity potential between the source and the sink is computed as 2.1042. The exact value is known to be 2.1126, and the error is 0.4 %. In the singularity programming, only the non-singular part is dealt with by a numerical scheme and the singular part is handled analytically. Even with a small number of $n_b = 8$, the CVBEM yields the results accurate enough for practical purposes.

References

1. Abramowitz, M., Stegun, I.A.: Handbook of Mathematical Functions with Formulas, Graphs, and Mathematical Tables. National Bureau of Standards, Washington (1970)
2. Bak, J., Newman, D.J.: Complex Analysis. Springer, New York (2010)
3. Bear, J.: Dynamics of Fluids in Porous Media. American Elsevier, New York (1972)
4. Brebbia, C.A., Telles, J.C.F., Wrobel, L.C.: Boundary Element Techniques—Theory and Applications in Engineering. Springer, New York (1984)
5. Carnahan, B., Luther, H.A., Wilkes, J.O.: Applied Numerical Methods. Wiley, New York (1969)
6. Churchill, R.V., Brown, J.W.: Complex Variables and Applications. McGraw-Hill, New York (1990)
7. Gelhar, L.W., Collins, M.A.: General analysis of longitudinal dispersion in nonuniform flow. Water Resour. Res. **7**, 1511–1521 (1971)
8. Greenberg, M.D.: Application of Green's Functions in Science and Engineering. Prentice Hall, Englewood Cliffs (1971)
9. Hildebrand, F.B.: Advanced Calculus for Applications. Prentice Hall, Englewood Cliffs (1976)
10. Howie, J.M.: Complex Analysis. Springer, London (2003)
11. Hromadka II, T.V., Lai, C.: The Complex Variable Boundary Element Method in Engineering Analysis. Springer, New York (1987)
12. Hromadka II, T.V., Whitley, R.J.: Advances in the Complex Variable Boundary Element Method. Springer, New York (1998)
13. Kellogg, O.D.: Foundations of Potential Theory. Springer, Berlin (1929)
14. Kreyszig, E.: Advanced Engineering Mathematics. Wiley, New York (2011)
15. Kwok, Y.K.: Applied Complex Variables for Scientists and Engineers. Cambridge University Press, Cambridge (1990)
16. Lang, S.: Complex Analysis. Springer, New York (1993)
17. Liggett, J.A., Liu P.L-F.: The Boundary Integral Equation Method for Porous Media Flow. George Allen & Unwin, London (1983).
18. Mathews, J.H., Howell, R.W.: Complex Analysis for Mathematics and Engineering. Jones & Bartlett Learning, Sudbury (2012)
19. Muskat, M.: Physical Principles of Oil Production. McGraw-Hill, New York (1950)
20. Needham, T.: Visual Complex Analysis. Oxford University Press, Oxford (2000)
21. Sato, K.: Complex Variable Boundary Element Method for Potential Flow. Baifu-kan, Tokyo (2003). (in Japanese)
22. Strack, O.D.L.: Groundwater Mechanics. Prentice-Hall, Englewood Cliffs (1989)

© Springer International Publishing Switzerland 2015
K. Sato, *Complex Analysis for Practical Engineering*,
DOI 10.1007/978-3-319-13063-7

Index

© Springer International Publishing Switzerland 2015
K. Sato, *Complex Analysis for Practical Engineering*,
DOI 10.1007/978-3-319-13063-7

Printed in the United States
By Bookmasters